DIFFERENT PLACES, DIFFERENT VOICES

Gender and development in Africa, Asia and Latin America

Edited by
Janet Henshall Momsen and Vivian Kinnaird

London and New York

First published 1993
by Routledge
11 New Fetter Lane, London EC4P 4EE

Simultaneously published in the USA and Canada
by Routledge
29 West 35th Street, New York, NY 10001

Reprinted 1995

Typeset in Baskerville by
J&L Composition Ltd, Filey, North Yorkshire
Printed and bound in Great Britain by
Mackays of Chatham PLC, Chatham, Kent

British Library Cataloguing in Publication Data
A catalogue record for this book is available
from the British Library

Library of Congress Cataloguing in Publication Data
A catalogue record for this book is available
from the Library of Congress

ISBN 0-415-07538-6 (hbk)
ISBN 0-415-07563-7 (pbk)

CONTENTS

FIGURES

TABLES

CONTRIBUTORS

Asmah Ahmad (Ph.D. Birmingham) is Lecturer in the Department of Geography, Universitat Kebangsaan Malaysia, Selangor.

Elizabeth Ardayfio-Schandorf (Ph.D. Birmingham), teaches in the Department of Geography and Resource Development at the University of Ghana, Legon.

Hazel Barrett (Ph.D.) is Senior Lecturer in Geography at Derbyshire College of Higher Education, England.

Jane Benton (Ph.D.) is Senior Lecturer in Geography at the University of Hertfordshire, England.

Angela Browne (Ph.D.) is Senior Lecturer in Geography at Coventry University, England.

Amriah Buang (Ph.D. Michigan) is Lecturer in the Department of Geography, Universitat Kebangsaan Malaysia, Selangor.

Sonia Alves Calio (Ph.D. candidate São Paulo) is a Researcher at the University of São Paulo, Brazil.

Nora Chiang (Ph.D. Hawaii) is Professor and Chair in the Department of Geography, National Taiwan University, Taipei, Taiwan.

K. Maudood Elahi (Ph.D. Dunelm) is Professor of Geography at Jahangirnagar University, Dhaka, Bangladesh

Peggy Fairbairn-Dunlop (Ph.D. Macquarie) is Lecturer in the Department of Agriculture, University of the South Pacific, Apia, Western Samoa.

B. Hyma (Ph.D. Pittsburg) is Associate Professor in the Department of Geography, University of Waterloo, Canada.

B. Folasade Iyun (Ph.D. Ghana) is Senior Lecturer in the Department of Geography, University of Ibadan, Nigeria.

Vivian Kinnaird (M.A. Waterloo) is Lecturer in the Department of Geography, University of Sunderland, England.

Nancy Davis Lewis (Ph.D. Berkeley) is Associate Professor, University of Hawaii, Honolulu and Secretary General of the Pacific Science Association.

Donny Meertens lectures in the Department of Social Geography, University of Amsterdam, The Netherlands.

Janet Momsen (Ph.D. London) is Professor in the Department of Geography, University of California, Davis and Chair of the International Geographical Union Commission on Gender and Geography.

Victoria M. Mwaka (Ph.D.) is Associate Professor, Department of Geography, Makere University, Uganda.

Perez Nyamwange is a Ph.D. student at York University, Toronto.

E.A. Oke (Ph.D. Kentucky) is Senior Lecturer in the Department of Sociology, University of Ibadan, Nigeria.

Elizabeth Oughton (M.Phil. Oxon.) is Lecturer in the Department of Agricultural Economics and Food Marketing, University of Newcastle upon Tyne.

Sarah Radcliffe (Ph.D. Liverpool) is Lecturer in the Department of Geography, Royal Holloway and Bedford New College, University of London, England.

Parvati Raghuram (M.A. Delhi) is currently a Ph.D. candidate in the Department of Geography, University of Newcastle upon Tyne.

Saraswati Raju (Ph.D. Jawaharlal Nehru University) is Lecturer in the Centre for the Study of Regional Development, Jawaharlal Nehru University, Delhi.

Yoga Rasanayagam (Ph.D. Cantab.) is Associate Professor in the Department of Geography, University of Colombo, Sri Lanka.

Colin Sage (Ph.D. Dunelm) is Lecturer in the Environment Section, Wye College, University of London.

Vidyamali Samarasinghe (Ph.D. Cantab.) is Associate Professor at the American University, Washington, D.C.

Janet Townsend (D.Phil. Oxon.) is Lecturer in Geography, Durham University, England.

Rameswari Varma (Ph.D.) is Lecturer in the Development Studies Unit, University of Mysore, India.

Anoja Wickramasinghe (Ph.D. Sheffield) is Lecturer in Geography at the University of Peradeniya, Sri Lanka.

PREFACE

This book consists of selected papers given at the Commonwealth Geographical Bureau Workshop held at the University of Newcastle upon Tyne in April, 1989. This was the first conference on gender and development for geographers. It brought together nearly one hundred academic geographers and other social scientists, faculty, postgraduate and undergraduate students, from over 40 countries. In addition to Commonwealth citizens, participants came from Sweden, Norway, the Netherlands, Iceland, Spain, Taiwan, Japan and the United States. The Workshop was co-sponsored by the International Geographical Union Study Group on Gender and Geography. We are grateful for financial assistance from the Commonwealth Foundation, the International Geographical Union Study Group on Gender and Geography, the British Council, the Ford Foundation, the Institute of British Geographers, the Canadian High Commission, the Province of Alberta and the University of Newcastle upon Tyne.

Most people who came to the meeting had been working on gender and development in isolation. For all of us coming together at Newcastle was a very exciting learning experience and the beginning of many new friendships and research collaborations. One of the most striking aspects was the distinctive approaches of people from the different countries. Clearly the most urgent research problems were location specific. Research methodologies also varied. It was felt that these spatial differences, displayed by the voices of geographers from developing countries, was one of the most interesting discoveries of the Newcastle Workshop. This volume focuses on this new regionalism.

Several papers on Caribbean topics were presented at the workshop. These have been included in another book entitled *Women and Change: a Pan-Caribbean Perspective* edited by Janet Momsen. Thus, the Caribbean is not represented in this volume. Non-Commonwealth areas are also under-represented because of the original organisation of the meeting.

The editors wish to thank all those who helped in the organisation of the workshop, especially Philip Courtenay, Secretary of the Commonwealth Geographical Bureau and Michael Eden and Dennis Dwyer, CGB Treasurers. We are grateful to our contributors from around the world for their patience in coping with the editing of the book, to our publisher, Tristan Palmer for his enthusiasm and encouragement, to Neil Purvis, Pat Juliusson and Anne Rooke for the artwork and to Dave Passmore for his editorial advice and overall support.

Part I
INTRODUCTION

1

GEOGRAPHY, GENDER AND DEVELOPMENT

Vivian Kinnaird and Janet Momsen

This book is a contribution to the growing number of volumes on gender and development written and compiled by geographers (Momsen and Townsend 1987; Brydon and Chant 1989; Momsen 1991). But more importantly it also addresses the call for theoretical and contextual agendas in our approach to, and analysis of, gender and development and feminist geography (Foord and Gregson 1986; Brydon and Chant 1989; Christopherson 1989; Townsend 1991). We address this latter concern by considering both 'unity' and 'diversity' as major themes of this book. The unity emerges from the commonalities among the contributors, virtually all of whom are geographers and citizens of the British Commonwealth. Consequently, they are reporting on field research which focuses on gender issues in developing countries within the context of British colonial discourse either as citizens of Britain or as citizens of former British colonies. For the former, as citizens of an erstwhile imperialist power, they are aware of the influence of colonialism and bring this knowledge to their approach to, and interpretation of, their work. For the latter, as researchers and academics in their own countries, they view their subjects from within the same institutional system – as subalterns relating and interpreting sub-altern voices.

This unity has inherent dangers. Spivak argues that the colonialist discourse reduces the subaltern to the role of 'a native informant for first world intellectuals interested in the voice of the Other' (Spivak 1988: 284). In addition, third world scholars writing about their working class fellow citizens may also take their own middle-class culture as a normative referent (Mohanty 1991) and, as feminist geographers, have an interpretation of feminism that is profoundly rooted in Western discourse. Yet, despite these dangers of polarisation, as geographers we know that we always have positionality which affects not only our choice of topic but also the intellectual 'baggage' we bring to our work. We seem to find it difficult to avoid the dominant discourses of patriarchal and hegemonic Western scholarship because of the institutional and conservative (Christopherson 1989) system through which we operate. Ironically, however, many of us have been working in isolation within our own segments of academe and have often been made to feel 'outsiders' (Glick 1986; Christopherson 1989).

Our theme of diversity is exemplified by the heterogeneity of the people we are studying and the explanatory importance of difference both in space and through time. Geographers delight in the difference between places, yet seek to provide a comparative analysis of space, place and pattern. By focusing on place in our studies of third world gender issues we are emphasising a geographical focus and a context-dependent creativity. This approach is closely aligned with the postmodernist reinterpretation of a politics of identity as politics of location. As such, the 'subaltern's position is not that of the exotic to be saved . . . her position is naturalised and reinscribed over and over again through the practices of locale and location' (Probyn 1990: 186). An emphasis on location and position highlights a concern with the relationships between different identities.

The recognition of place-consciousness led us to group the papers within the book by continent. Each section is introduced by an overview written by feminist geographers working within that particular region. The introductions provide a very personal account of the position of geography within gender studies at universities in the respective regions and of the individual authors' experience with university geography departments. It is clear that feminist geography is stifled by a rigid definition of geography in many university departments.

The regional overviews are particularly useful for their interpretation of 'gender and geography' from different viewpoints. From the perspective of African, Asian or Latin American feminist geographers it is clear that, 'gender and geography' and 'gender and development' are one and the same concerns; a perspective somewhat different from Western geographers' scholarly appreciation of 'gender and development' as a subset of feminist geography and/or development studies concerned with 'others over there' and theoretically unpacking what we mean by 'development'. And yet, all these feminist geographers are working within a (feminist) geographical academic discourse that has been shaped and defined by Western scholarship. As feminist geographers, our intellectual training is filtered through our own cultural knowledge and is constrained by our position in society. As such, our experience of geographical enquiry, our research methodology and our regional viewpoints are dependent upon our knowledge creation as individuals exposed to different intellectual stimulation and our position *vis-à-vis* different concerns and issues, both conceptually and empirically. This distinctiveness was a major learning experience for all of us who attended the conference upon which this book is based. It gave the various papers a contextual unity, an embeddedness within different distinct localities, in each continental region.

In celebrating the context-specific issues of individual 'regions', we are seeking to do more than create a spatial representation of important continental concerns. The United Nations (1991) has noted regional patterns in the global trends affecting women's position over the last two decades and in the problems they face today. In Latin America, rural/urban differences have been exacerbated as urban women have gained greater access to health and education as well as a higher degree of economic and political participation. In sub-Saharan Africa there has been very limited progress in reducing gender inequality in most social, economic and political contexts. Moreover, the general economic decline that has become characteristic of the 1980s would suggest a continuation of pressure on living standards and resource use for both women and men. Women in northern Africa and western Asia continue to lag far behind other regions in terms of their participation in social and economic activities and political decision making. In south Asia, women's health and education levels have seen some improvement but remain greatly inferior to those of men, while the benefits of economic growth have rarely accrued to women. In south-eastern and eastern Asia and the Pacific, urban/rural difference in standard of living and gender inequality have declined but women are still excluded from most decision-making processes. While useful in a general sense, these descriptions merely transform gender differences into the division between women and men (Mohanty 1991) rendering an analysis of the particulars of locality from both contemporary and historical perspectives very difficult indeed.

Clearly the context within which women deal with contemporary crises and the nature of the crises themselves varies from region to region. However, we see women as agents of change responding to the crises facing their societies in the modern world. While specific crises may be attributable to specific regions, for example the impact of structural adjustment in Latin America or the extreme environmental degradation of sub-Saharan Africa, generalised statistical data, as reported in the United Nations' document, provides insufficient evidence for conceptualising women's response to these crises. Indeed, constructive conceptualisation and theory building can only come from empirical research that pays specific attention to those conditions which help explain gender relations and, for example, the reasons *why* women's health in south Asia is inferior to that of men.

It is to this particular concern that the papers of this volume are addressed. While 'regional' themes are frequently apparent, all papers focus on specific localities and contexts. In Africa, a strong theme is the environment in rural areas. Ardayfio-Schandorf (Chapter 2) analyses the different experiences of women in three contrasting villages in the savanna and forest ecozones of Ghana. Her concern focuses on the constraints women face in the provision of rural energy supplies. Although all of the women in her study face a similar (regional) problem, their experience and interaction with the environment differs from one location to another and finding solutions to the problem of energy provision requires location specific consideration. Similarly, Hyma and Nyamwange (Chapter 3) focus on the problem of resource depletion in Kiambu, Kenya. Their careful consideration of the resource management issues, the knowledge and practice of women in forestry management and the institutional framework within which the entire system operates highlights the way in which women's participation in environmental initiatives in order to restore forest cover can lead to a strong programme for sustainable development. Mwaka (Chapter 4) suggests that the political economy of environmental degradation is at least partially responsible for the decline in agricultural production in Uganda.

In Latin America, we see a focus on rural/urban interaction and rural/rural interaction through migration. The papers by Meertens (Chapter 19), Townsend (Chapter 20) and Radcliffe (Chapter 21) look at the changing gender relations in the migration process. Meertens and Townsend consider the changing nature of women's social and economic position as they and their families migrate to frontier settlements. Townsend's case study in the Middle Magdelena Valley and Meerten's case study of the San Jose del Guaviare, highlight women's different experience in the process of frontier colonisation in Colombia.

In south Asia, the focus was on the interaction of women's productive and reproductive roles and decision making especially in rural areas. The strength of patriarchal gender relations in many of these case studies was particularly convincing. Elahi (Chapter 7) provides evidence to account for differences between male and female health and mortality in two villages in Bangladesh. Society's preference for male children, cultural norms regarding the young age at which women marry, and the differential access that men and women have to the decision-making process regarding health practices help to explain the 'official statistics'. Varma (Chapter 10) and Rasanayagam (Chapter 12) focus on male bias in the rural development process in southern India and Sri Lanka, respectively. Both find that women's participation rate as agents and beneficiaries of rural development programmes is fairly low. Wickramasinghe (Chapter 13) and Samarasinghe (Chapter 11) analyse women's experience in rural Sri Lanka. Samarasinghe highlights the experience of migrant Indian Tamil tea plantation workers suggesting that their quality of life is lower than that of rural plantation workers as a whole. This distinction owes much to the genesis of the plantation structure itself, which separates immigrant labourers from the rest of the population. Both authors utilise time budget surveys to decipher women's commitment to productive and reproductive work. Finally Oughton (Chapter 8) and Raghuram (Chapter 9) use field surveys from west India to provide a detailed analysis of the seasonality of the gender division of labour in a rural village. The contribution of women's wage and non-wage labour are both vital to the household and to the agricultural production process.

In south-eastern Asia and the Pacific, all the papers looked at the confrontation between traditional ways (religious, cultural) and modern Western influences both in the urban–industrial context (Malaysia) and in the rural context (Western Samoa). Buang (Chapter 15) reaffirms earlier work on the social impact of Western multinational 'culture' among Islamic communities in Malaysia. Ahmad (Chapter 14) is concerned with women's perception of life aboard raft-houses in a peri-urban environment. Close proximity to the city and its employment opportunities are juxtaposed against a feeling of 'ruralness' offered by the raft-house community. Fairbairn-Dunlop (Chapter 16) expresses the strength of cultural norms in Western Samoa to explain women's lack of access to agricultural development and modernisation.

Yet, these themes do not address all of the papers in this collection. Hazel Barrett and Angela Browne (Chapter 5) analyse the impact of *coos* mills for village women in The Gambia. Bose Iyun and E. A. Oke (Chapter 6) are concerned with urban market women's knowledge and use of family planning services and find that women's place of residence (and therefore work) has an impact on their knowledge and use of contraception in an urban setting. Jane Benton (Chapter 17) evaluates the significance of women's involvement in non-governmental organisations in urban Boliva and their power to engage in activities which bring about positive changes for women, their families and their communities. Colin Sage (Chapter 18) explores the role and importance of women's economic activities in relation to the organisational structure and resource base of the household and illustrates the importance of understanding the dynamics of individual lives in rural Bolivia.

Many of these 'topics' fall firmly within established geographical fields of enquiry: human/environment relations, demographic analyses and migration. Others exemplify the broad range of issues as part of a 'new geography' that is bold in embracing new areas of enquiry and new methodologies.

The papers included in this book were presented at the first conference of the International Geographical Union Study Group on Gender and Geography. The excitement felt during the meeting at Newcastle upon Tyne in 1989 has continued. Today the Group has over 350 members in some 50 countries worldwide and became a full-fledged

Commission in August 1992. It has co-sponsored meetings in Brazil, Hawaii, China and England and in 1992 held its second conference, this time in the United States. The group has a regular newsletter through which research interaction is encouraged and publishes a series of Working Papers. This volume forms the first compendium of research by members of the study group on issues concerned with gender and development. It has already triggered many new research projects, the fruits of which will soon enlarge the hitherto limited contribution of geographers to this field.

Part II
AFRICA

INTRODUCTION

Gender and geography in Africa

Elizabeth Ardayfio-Schandorf

The situation of gender and geography in Africa can be described as heterogeneous and fragmentary. Even though geography, as a field of study, is holistic in its approach, in the traditional discipline, little attention has been paid to women's issues or gender relations until recently. This situation is even worse in Africa where geographers have taken a much longer time to join hands with other disciplines in the quest for knowledge and understanding of women and gender issues. Earlier scholarly work had been contributions in social and economic geography in fields such as historical, population and rural development where women's issues were marginally or incidentally studied. Some of these studies merely pronounced the presence of women and others to document their important contribution in various spheres of national development.

With the wind of change that swept across the globe through the launching of the International Women's Year in 1975 by the United Nations, geographers in Africa have been increasingly drawn into feminist studies which have proliferated elsewhere in the world as a multidisciplinary field. The momentum in the developed countries, particularly the West, has given birth to women's studies and women's studies scholars. Such scholars have *inter alia* developed appropriate theories, concepts and assumptions on how to construct gender and conceptualise women's issues. By doing so they have shed deeper insight into the situation of women.

These developments have had great influence on African feminist scholarship. The interest in women's studies as a separate discipline is being increasingly recognised in African universities. For instance, Makerere University of Uganda has recently established a women's studies department. A research and documentation centre has been established at the University of Ibadan, Nigeria. An increasing interest in these types of institutions is evident.

Just like their counterparts in the arts and social sciences, geographers' contributions to women's studies are demonstrated through research initiatives and curricula development. In the field of academic research, issues of patriarchy and power relations, class, race and ethnicity, have failed to be strong foci of enquiry. This is partly due to the lack of a comprehensive framework for analysing gender in geographical work, and the strong patriarchal relations that prevail in African societies. Most known geography scholars are undertaking academic research on women and gender in both Anglophone and Francophone Africa in such areas as agriculture, including fishing; environmental resource utilisation and management; demography; and employment of women in both rural and urban areas. Most of these studies have stemmed from academic and development research by geographers who are contributing to equitable development by documenting and analysing the situation of women nationally, and regionally, to attract attention to the position of women by indicating areas where policy intervention may be needed.

The work of geographers on women seems to point in one major direction, despite the complexities perceived through class and gender analysis in women's studies. The environmental crisis in the whole of Africa, with its persistent drought and desertification particularly in northern and western Africa, has further heightened interest among geographers. This trend has been reinforced by the global concern for the environment as demonstrated by the Bruntland Report (World Commission on Environment and Development 1987) and the African Ministerial Conference on the Environment in Cairo (1985). Presently, a clear emphasis of geographic research is on management of the environment and environmental studies relating to soil, water, forests, and renewable energy resources. This is an area where scholars in geography incorporate gender and thereby contribute positively to development research. Dankleman and Davidson's edited volume *Women and Environment in the Third World: Alliance for the Future* and Ardayfio-Schandorf's work on women and the rural energy crisis are contributions which draw attention to the environmental crisis in Africa and its effect on women's work, nutrition and family welfare. This direction is confirmed by the 1989 IGU Workshop on Geography and Gender, where 70 per cent of the papers presented pertaining to Africa, related to an environmental theme.

Current research initiatives also have a strong leaning toward environmental issues. Women, Environment, Development Network (WEDNET) seeks to draw African women into a network to develop conceptual frameworks and methodologies for studying women and environment in Africa. Within this initiative, 'Women and Natural Resource Management' is funded by the International Development Research Council (IDRC) and is managed by the Environment Liaison Centre International (ELCI) and seeks to legitimise women's environmental knowledge and innovative strategies in the promotion of sustainable development.

This on-going work and research may lead to new ways in women's research as they couple practical knowledge with theoretical approaches. Questions of equity, regional integration and peace need to be asked and may influence the future direction of gender and geography in Africa.

This direction will strengthen gender and geography as a separate field of study within Africa. Eventually, the accumulation of scholarly work on women and environment may pave a way for the development of a paradigm focusing on sustainable utilisation of resources through a people-centred approach to development. The emerging conceptual approach must be based on development concepts which are ecologically harmonious because of the intimate relationship between the productive and reproductive functions of women and the environment and its implications for health, nutrition, and political participation.

To develop such a new paradigm, the availability of both qualitative and reliable quantitative data is highly desirable. Statistical data disaggregated by sex are already in the process of being further developed by national governments as well as United Nations' specialised agencies. More empirical research by geographers, on the basis of gender and class, will yield more information while the quantitative data will provide a basis for more precise analysis of the trends. In addition to national and international statistics of socioeconomic indicators of development, various categories of women and men in urban and rural settings need to be studied. In order to do this some of the parameters that could be taken into consideration are employment status, gender relations over time to reflect the occupation structure, settlement patterns, population and family planning, and spatial mobility. The different types of migration and their impact on both women and men should equally be studied. Access to the various levels of education and the professions and discrimination against women in the formal labour market are other important areas of research.

In the rural context, the critical role of environmental and agricultural factors and their developmental relationships, including agrarian structures and reforms need to be examined. As land is crucial to relations of production in African rural economies, research and studies on land including land rights, ownership and distribution by gender should also be included. Indeed, the land use pattern including the various cropping systems is also important. This approach will help identify the marginalisation of the rural poor including women so that appropriate planning tools will be devised for promoting effective social change. Studies should be undertaken of the disparities in the development of these relationships at the district, inter- and intra-regional as well as sub-regional levels due to the current processes of decentralisation in African countries. This will help to determine development patterns and cultural lineaments.

These patterns of development should also be related to the strategies of national machineries, such as the National Council of Women and Development (NCWD) of Ghana. Even more importantly, gender should be considered in the context of African governments' policies on structural economic transformation processes. As the patterns and trends unfold, areas where resources should be redirected and re-allocated will become much clearer. Policy studies could then direct those resources to the deprived areas or regions for allocation to vulnerable groups for more equitable development.

2

HOUSEHOLD ENERGY SUPPLY AND RURAL WOMEN'S WORK IN GHANA

Elizabeth Ardayfio-Schandorf[1]

INTRODUCTION

Environmental degradation is putting increasing pressure on Ghanaian rural women. The household economy in which rural women actively participate is based almost entirely on biomass. The extent of their dependence on environmental resources is clearly demonstrated through the income they derive from agricultural produce and other small-scale economic enterprise. These monetised activities as well as non-monetised activities depend on reliable rural energy supplies. In this respect changes in environmental conditions are a threat to the economic survival of women, particularly for those in the savanna agro-ecological area. Their problems are exacerbated by the process of development in the country. In this chapter, the constraints women face in the provision of rural fuel supplies, in three contrasting villages, are considered in relation to women's productive and non-productive activities in the forest and savanna ecozones.

> Though woodfuel is in principle a renewable resource, the fact is that in many parts of the world it is being depleted just as surely as if it were an irreplaceable fossil fuel. If present trends continue, the woodfuel supplies of many hundreds of millions of people will be exhausted long before the oil fields on which the industrialized world depends have run dry.
>
> (Barnard 1987)

Recent research results indicate that 80 per cent of all households in the developing countries depend on wood as their primary source of energy. Global consumption of firewood is much higher than that of energy from hydro-power, nuclear power and geothermal sources put together. Fossil fuels and other alternative fuels such as natural gas and biogas, have not brought about any appreciable impact on energy consumption in Africa at the household level. Fuelwood is needed in Ghana, as in other developing countries for mechanical power and for survival tasks.

The high cost of oil imports constrains the nation's ability to sustain its development strategies (Ardayfio 1983). Imports of petroleum cost 41 per cent of Ghana's foreign exchange in 1981, and over 60 per cent of the nation's total earnings were spent on petroleum and its by-products in 1983 (Ardayfio 1986). Woodfuel and hydro-power are the most important energy sources, constituting 86 per cent of total Ghanaian consumption in 1987. Woodfuel alone accounted for 92 per cent while hydro-power comprised 8 per cent of the domestic energy output (Ghana Government, Ministry of Fuel and Power 1988).

The Forestry Department may not identify scarcity now, (Barnor 1985) but, according to FAO estimates (1986), by the year 2010, Ghana will experience a deficit. It is among those third world countries which are 'consuming more than permissible amounts of wood' (Parikh 1980). Heavy demand for fuelwood and charcoal coupled with unreliable supplies of fossil fuel and the rising price of hydro-electric power, have deleterious consequences for the environment. Population growth puts pressure on forest resources, and such pressure may bring about changing patterns in rural energy consumption patterns. To improve energy supplies, alternative energy sources should replace traditional woodfuel resources. In order to do this it is imperative to understand fuelwood production and consumption patterns, by producing reliable data in order to demonstrate the possible impact on the environment and human activities.

THE ENERGY RESOURCE BASE IN GHANA

Energy used by Ghanaian society is derived basically from biomass resources. This usage pattern is clearly related to the ecology and the social mode of production of the local communities. Ecologically, two major regions, namely the forest zone, embracing about a third of the country and the savanna zone covering the remaining two thirds of the country can be identified. The forest area is more prosperous and contains about 66 per cent of the total population of the country. Wood resources diminish progressively northward through the derived savanna, the southern savanna, and the northern Guinea, and Sudan savannas. Within this zone cash crop production gives way to domestic crop production and animal husbandry due to increasing arid conditions.

The forests provide the main sources of timber and logs for export and generally for the wood industry. Fuelwood and charcoal are derived from these sources as well as from the savanna woodland, which also yields valuable poles for the local building industry. As the area under agriculture expands, deforestation is hastened, thus limiting the availability of woodfuels. According to the land use map of the country only 2 per cent of the limited forests, which are unreserved, are available for cultivation.

A balanced pattern of land use and environmental exploitation is important in maintaining the resource base on which rural energy supplies depend. Industries such as those of timber, plywood and food also exert tremendous pressure on the forests and cultivated land. Crop lands, forests and energy needs suffer as a result of environmental degradation caused by the high demand from the metropolitan powers for timber products, and export crops (like cocoa whose production has recently spread to the south-western part of the country). It is, thus, the external demand generated by these products (Agarwal 1985) that has contributed to environmental degradation within the forest areas. Added to this, is the expansion of local agricultural production which has been necessitated by rapid population growth at a rate of 2.6 per cent to 3 per cent per annum. The fallow period has had to be shortened resulting in lower agricultural productivity. In the savanna areas the introduction of mechanised land clearing has been a contributory factor in the destruction of trees and soil erosion in the less wooded areas exacerbated by climatic change in the sub-region with the worst drought occurring in the 1960s. Industrial establishments and urban sprawl also encroach on the wooded and forested zones in peri-urban parts of the country.

By 1970 there was a shift in the urban areas from the use of fuelwood to charcoal with the percentage using charcoal for cooking increasing from 43 per cent to 69 per cent. This trend is more wasteful of wood resources (Ghana Population Census 1970) as increased urban demand for wood and charcoal exacerbates the environmental problem.

WOMEN AND CHANGING RURAL ENERGY SUPPLIES

The ecosystem has been undergoing various changes which have become more conspicuous over the past three decades. Particularly during the period between 1969 and 1983, the natural vegetation underwent rapid deterioration. Extensive areas of high forest have been degraded into secondary forests while the latter have also been colonised by grass in the marginal areas. Moreover, desertification has occurred in the savanna zone as the rains have become unreliable and unpredictable. Rainfall has tended to be erratic and of short duration, reducing mean annual precipitation values. The severe droughts which seemed to be unprecedented in living memory, impoverished the vegetation and paved the way for devastating bush fires that destroyed vegetal cover, crops and wildlife as well as human beings.

It is clear from the conclusions of the Ministerial Conference on Drought and Desertification in Africa that changes in the vegetation cover have been caused by over-cultivation and expansion of agricultural land. Now the level of fuelwood demand exceeds supply but because of a paucity of data on the proportion of fuelwood actually consumed in rural areas, it is difficult to apportion the growth in per capita fuelwood consumption between population movement to the urban areas and the shift from wood to charcoal in urban households. It has been suggested that the apparent growth in fuelwood utilisation is due to the importance of fuelwood for commercial purposes. It is increasingly used in artisanal activities, such as the preparation of beverages, palm oil or soap. In northern Ghana, the situation has gone a stage further, *shea* butter processing, for example, used to be one of the most widespread activities and also the most energy intensive of all the rural industries in the area. With deterioration in the vegetation and the spread of desertification, wood resources have become so scarce that many women have been compelled to withdraw from this industry.

The process of modernisation has also tended to limit women's access to and supply of household resources. Colonialism introduced an institutionalised system which transformed the economic behaviour of Ghanaian women and men. Men were encouraged to cultivate export crops for the metropolitan markets using their own lands. Those with less opportunity to engage in such agricultural production migrated to the cocoa growing areas to sell their services as labourers on the cash crop farms. As more and more men went into cash crop production, women, who were discouraged from participating in the agricultural export economy, were narrowly restricted to the production of subsistence crops for feeding the family (Boserup 1970). Through this new system, 'inequality between men and women (under colonialism) emerged as an aspect of social differentiation characteristic of capitalistic society where sharp differences exist between persons on the basis of unequal access to the means of production' (Pala and Seidman 1976).

As Ghanaian men were required to enter into productive work they were favoured in their access to productive resources as new attitudes undermined the original status of women. This could be seen in women's limited access to land and their capacity to manage this critical resource which is a key factor of development in rural Ghanaian society, just as in all traditional African societies (Udo 1982; Pogucki 1955; Bukh 1979).

In the process, the attention of men was shifted from domestic production (which became the lot of women) to cash and export crop production. In the rural political economy, therefore, women clear the undergrowth, collect dug-out tree stumps and roots during farm-land preparation, apply fertiliser, weed round the farms, sow seeds and harvest vegetables (Ardayfio 1986). According to the 1984 national population census, in food crop production and animal husbandry, women outnumbered men, comprising 51 per cent of all people engaged in agriculture. Women also assist husbands to maintain cash-crop farms by weeding and harvesting. Thus, the pattern of development in the country has tended to reduce the relative value of the economic contribution of women. Since they lack education and the skills required in modern society, as advances like the introduction of modern techniques are made in agriculture, they tend to be by-passed by technical developments. Women normally own smaller acreages than men, thus, limiting the use of modern equipment and their access to credit (Okali and Kotey 1975).

Privatisation of land not only broke down the land tenure system but also the land-holding and land-use pattern. As a result, fragmentation of agricultural land developed with population pressure and the need to share land among family members and children from one generation to another. The impact of this system on women and rural labour is enormous. Women as primary food producers are faced with numerous challenges as they organise themselves in the rural milieu. The basic challenges include provision of rural energy and water and making decisions according to their own interests in order to enhance their capacity for self-reliance in productive activity, earn more income and improve their quality of life. As women are deprived of the basic means of production, they organise and adopt strategies that will reap certain benefits from the system, given their intimate knowledge of the environment for survival.

WOMEN'S ECONOMIC ENTERPRISE

Generally, women's work involves both monetised and non-monetised tasks both of which are crucial in household survival and rural development and are heavily dependent on rural energy supply. In the field of industry and manufacturing they are engaged in small-scale production such as food processing and charcoal making with fuelwood as a very important input. Through the performance of such monetised tasks, women contribute more to the household economy than their male counterparts. From the results obtained from the survey of three villages in the forest and savanna ecosystems, it was clearly demonstrated that women's real contribution to the economic and social life of the country deserves more attention than hitherto acknowledged.

In the fishing village, during the first rainy season survey, for example, women contributed as much as 68 per cent of the total household income, primarily earned from the smoking and distribution of fish. However, by the dry season, women's contribution to household income had increased to 90 per cent. It appears, however, that the contribution of men's income is generally not as low as the survey suggests. The survey findings could be attributed, to a large extent, to the failure of food crops for the season, due to the exceptional drought of 1983.

In the forest village also, the income contribution of women exceeded that of men ranging from 60 to 63 per cent in the rainy and dry seasons, respectively. Their income was derived from the sale of staples and cocoyam and from food processing of such items as palm oil and gari.

The situation was quite different in the savanna village where, during the first survey, men's contribution was higher providing 55 per cent of the total household income. Their sources of income were mostly from the sale of vegetables and other forms of work, whereas women only sold vegetables. By the second survey, women's contribution had superseded that of men, providing 65 per cent of the household income. When agricultural produce did not yield any income, charcoal manufacturing became the major source of income.

The high contribution of women's income could be attributed to the fact that women in all the villages had at least two sources of income. The major source was either farming or fishing, depending on whether the settlement was located along the coast or in the hinterland. Thus, in the savanna and forest villages, the women were mainly farmers and their secondary source of income was food processing. During the dry season women's primary activity in the forest was food processing, while in the savanna village, most women shifted to charcoal production due to high demand for that commodity in the nearby urban areas. In the fishing village, on the other hand, women undertook food processing or petty trading as a means of supplementing their major sources of income. Indeed, the ability of women to have a subsidiary source of revenue goes a long way in explaining their higher contribution than men to overall household income. This dominance becomes even more apparent when the major sources of income, closely related as they are to the climate, bring below-average amounts.

In their farming operations, women cultivate their own individual plots. Over the past ten years, the acreages of both family farms and women's individual plots have been contracting, even though the same types of crops continue to be cultivated. Women provide their own labour, supplemented by that of household members and sometimes by other members of the extended family. In all, women's and other household members' labour constitutes 77.8 per cent of labour on women's farms with hired help constituting only 22.2 per cent. Non-family support is employed during the clearing or peak season.

Women's contribution to trade is related to the agricultural produce they cultivate ranging from staples to vegetables and tree crops such as oranges, avocado pears and palm oil. In addition, women trade in other agricultural produce, such as yams and onions, purchased from Accra for retailing in their communities. Such women also trade in manufactured items that are locally produced or imported.

Besides farming and trade, small-scale industries provide women with income. In 27 per cent of the households surveyed, such enterprises were based on natural resources, while those of men, where they existed, appeared to be more oriented towards the 'modern' sector. Small-scale industries fall into two main groups: food processing and non-food manufacturing. The food-processing industry is a major source of income to women as food that is consumed in urban areas is generally produced and processed in the rural centres. The energy component is essential to these industries but 'its complex trends are inadequately recognised and, therefore, it does not receive the attention it deserves' (Ghai 1986).

In the surveyed areas small-scale food processing for income includes the processing of corn, cassava, palm oil, fish, rice and vegetables. Almost all women are engaged in some form of processing activity either as a major or secondary occupation. Most of these industries are energy intensive and rather inefficient. In work undertaken in Ghana and Sierra Leone, it was found that in the case of fish smoking, fuel use per kilogram in large traditional ovens is relatively low, but most women use the small ovens which use 0.125 kg of fuelwood per 1 kg of fish. In the processing of palm oil, fuelwood is used and on a three-stone fire, the average fuel use is high, consuming 7 kg per gallon of oil. Similarly, rice cooking is fuel intensive using on average 3.9 kg per bushel of rice (Stevens 1986).

WOMEN IN RURAL ENERGY SYSTEMS

As household management is the traditional responsibility of women, the provision of energy for cooking is a central task as women have to ensure that there is always adequate energy for household requirements. But with deterioration in the environment, the task of procuring fuelwood is increasingly difficult (Table 2.1). In the fishing village, for instance, 95 per cent of women did not have any difficulty in securing fuelwood 10 years ago but now this is true for only 5 per cent. In addition, women experience shortages in the supply of good wood species on their farms.

LABOUR ALLOCATION OF WOMEN

The increasing problems of fuelwood supply lead to conflicts for women who combine their role in productive activities such as farming, fishmongering and market gardening with domestic activities. In order to

Table 2.1 Problems encountered in fuelwood production

	Savanna village		*Fishing village*		*Forest village*	
Problems	*Now* %	*10 yrs ago* %	*Now* %	*10yrs ago* %	*Now* %	*10yrs ago* %
Inadequate supply	38.1	28.6	20.0	0	28.6	0
More expensive	0	0	40.0	0	0	0
Walk further distance	38.1	14.2	25.0	0	9.5	14.3
Diminished good species	4.8	4.8	10.0	0	0	0
No problem	4.8	48.0	5.0	95.0	42.9	47.6
No answer	14.3	4.8	0	0	14.3	9.5
Not applicable	0	0	0	5.0	4.8	28.6
Total	100.0[a]	100.0[a]	100.0	100.0	100.0[a]	100.0

Source: Household survey, 1983
Note: [a] Some totals are rounded

investigate how their coping strategies are affected, the household economy including energy provision and all other interrelated factors and activities in which women participate is considered. In the areas studied, the participation of women in these activities varied from one village to another, depending on the location, the nature of the local economy and the agro-ecological zonation of the settlement. During the 14-hour day most women spent more time on domestic activities than on income-generating activities. In the savanna village, female farmers and traders spent, on average, 3.5 hours in economic activities with the remaining 10.5 used for mainly domestic activities. The time spent by women in these latter activities is enormous, due to the need to stock fuelwood during the farm-clearing season. In the forest ecosystem, most households produce only 10–19 kg of wood daily, whereas in the

Table 2.2 Daily quantities of fuelwood produced (minor farming period)

	Households		
Quantity kg	*Savanna village* %	*Fishing village* %	*Forest village* %
1–9	14.3	9.1	9.1
10–19	14.3	9.1	57.1
20–29	42.8	9.1	28.6
30–39	28.6	45.5	14.3
40 and above	0	27.2	0
Total	100.0	100.0	100.0[a]

Source: Household survey, June, 1983
Note: [a] Some totals are rounded

Table 2.3 Daily quantities of fuelwood produced (major farming period)

	Households		
Quantity kg	*Savanna village* %	*Fishing village* %	*Forest village* %
1–9	0	0	16.7
10–19	16.7	16.7	25.0
20–29	33.3	50.6	50.0
30–39	50.0	16.7	8.3
40 and over	0	16.7	0
Total	100.0	100.0[a]	100.0

Source: Household survey, August, 1983
Note: [a] Some totals are rounded

savanna, households tend to produce between 20–39 kg of fuelwood a day during both seasons (Tables 2.2 and 2.3). In the latter, where fuelwood production is problematic, a fuelwood procurement trip is an opportunity for the household to bring in as much fuelwood as possible to cut down on the labour involved in the process.

In the fishing village, women spend about 6.3 hours on income-generating tasks, mainly fishmongering, while in the forest women spend approximately two hours per day on income-generating activities. Normally, the hours women devote to economic activities would be higher than recorded here due to the circumstances prevailing at the time of the survey (a long period of drought, crop failures and bush fires). During the second survey, which took place during what was supposed to be a minor harvesting season, there was no harvest and some of the women directed their attention to charcoal making, fuelwood collection and house cleaning. In the fishing village where the number of hours recorded as being spent on income-generating activities was much higher than in the two other villages, the survey coincided with the beginning of the fishing season when fish smoking demanded the attention of the women.

As regards domestic activities, a woman farmer in a savanna village spends on average 1.3 hours per day on water collection alone. This follows food processing and cooking which together take 2.1 hours of her time. The third activity is fuelwood production which takes 0.8 of an hour of her day, followed by child care. Clearly, the major task is cooking, taking about 2 hours including preparation time. Female traders spend less time in fuel collection because they operate between the village and city markets and prefer to purchase fuel.

In the fishing village, women farmers spent two hours in cooking per day (excluding food preparation), whereas the fishmongers spend only 0.6 of an hour cooking. Moreover, fishmongers are not directly involved with fuel and water collection and their domestic tasks centre on cleaning the compound which is related to their fish smoking business.

During the May–June period, most women cooked only one meal per day, mainly in the evening, with a few also preparing porridge in the morning. In some villages, the situation had deteriorated to starvation level by the second survey in August, which would normally fall within the normal harvesting season. In the two savanna villages, the food situation was so grave that women spent less time cooking. Their observed labour-allocation time was 1.5 and 2.4 hours respectively. Indeed, there were variations in cooking time throughout the year and also in the varieties of food being cooked.

In the forest village, as food supplies improved, the time allocated to cooking also increased to 6.5 hours a day, and 6 hours in subsequent surveys. This more truly reflects the time women normally devote to cooking in West African countries. In a similar study in Nigeria, it was found that women in Yorubaland spend on average between 4 and 6 hours per day in the preparation of meals (Ardayfio-Schandorf 1983).

The amount of time generally devoted by women to cooking is related to their other activities. In the fishing village, women involved in fish smoking were observed to spend on average 9 hours per day on that task and less than 2 hours per day on cooking. In certain instances, as in the fishing village, there were days when women did no cooking at all. In that case, other household activities took about three hours. Similarly, retail traders, charcoal manufacturers and commercial fuelwood producers spent less time in cooking and housework than their counterparts in other forms of work. Excluding fuelwood production, it was observed that women took between 6 and 7 hours per day to produce charcoal but the actual time input was about 10 to 12 hours, depending on the quantity of wood being converted to charcoal.

Table 2.4 Distance travelled in household fuelwood production (minor farming period)

	Households		
Distance km	*Savanna* %	*Fishing* %	*Forest* %
1–2	57.1	0	28.6
3–4	28.5	18.2	7.1
5–6	0	0	28.6
7–8	0	9.1	7.1
9–10	14.3	0	21.4
11+	0	72.7	7.1
Total	100.0[a]	100.0	100.0[a]

Source: Household survey, June, 1983
Note: [a] Some totals are rounded

Another time-consuming household task is fuelwood production. The female labour involved in fuelwood production includes pollarding, cutting, splitting and gathering of wood and also travel to the production sites. In the savanna woodland mosaic zone, among the women of the fishing village, most walk for more than 11 kilometres to obtain fuelwood (Table 2.4). Even in the forest ecosystem where wood is relatively plentiful, women sometimes walk more than 11 kilometres in order to get to a production site (Table 2.5). In this area, women make longer trips because their farms are scattered due to the rugged nature of the terrain. They have to walk beyond the hills to get to the fuelwood site. In addition to walking time, the actual collection of wood is time consuming.

One would expect production time to increase as fuelwood availability declines, but this is not always the case. During the first survey at the time of the minor clearing period, it took most women of the savanna villages over 45 minutes (Table 2.6) to produce fuelwood. However,

Table 2.5 Distance travelled in household fuelwood production (major farming period)

	Households		
Distance km	*Savanna* %	*Fishing* %	*Forest* %
1–2	33.3	66.7	16.7
3–4	16.7	33.3	0
5–6	50.0	0	33.3
7–8	0	0	0
9–10	0	0	41.7
11+	0	0	8.3
Total	100.0	100.0	100.0

Source: Household survey, August, 1983

Table 2.6 Time spent in household fuelwood production (minor farming period)

	Households		
Time taken (minutes)	*Savanna* %	*Fishing* %	*Forest* %
1–15	0	0	21.4
16–30	28.6	36.4	42.9
31–45	0	9.1	21.4
46–60	42.9	45.5	14.3
60+	28.6	9.1	0
Total	100.0[a]	100.0[a]	100.0

Source: Household survey, June, 1983
Note: [a] Some totals are rounded

Table 2.7 Time spent in household fuelwood production (major farming period)

Time taken (minutes)	*Households* Savanna %	Fishing %	Forest %
1–15	0	0	16.7
16–30	33.3	50.0	25.0
31–45	50.0	33.3	8.3
46–60	16.7	16.7	8.3
60+	0	0	41.7
Total	100.0	100.0	100.0

Source: Household survey, 1983

during the major clearing season it took about half of the women interviewed up to 60 minutes to obtain fuelwood (Table 2.7).

The high labour and time demands of fuelwood collection create a triple burden for women on top of their productive and reproductive roles. By way of relief their household members may be employed in fuelwood production. This occurs most often during the major producing season in June when children and husbands assist in production.

On average, children provided 15 per cent of the labour at this time (Tables 2.8 and 2.9). By the second survey, however, the contribution of all household members, including that of children, had dropped in contrast to that of women which had risen sharply to 79.2 per cent. Women, together with children, provided over 60 per cent of the labour input in all three villages during the first survey with their participation rising to 95 per cent during the second survey. Thus, fuel production remains an important task for rural women. Nevertheless, men now tend to participate more often in fuel production to extents that vary from village to village. In the fishing village, men were observed carrying fuelwood. In the savanna village, men were increasingly involved in

Table 2.8 Distribution of labour within households for fuelwood production (minor farming period)

Who	*% of labour* Savanna	Fishing	Forest
Women	28.6	18.2	64.3
Men	42.9	0	0
Children	14.3	18.2	14.3
Women and children	0	63.6	21.4
Other	14.3	0	0
Total	100.0[a]	100.0	100.0

Source: Household survey, June, 1983
Note: Some totals are rounded

Table 2.9 Distribution of labour within households for fuelwood production (major farming period)

	% of labour		
Who	*Savanna*	*Fishing*	*Forest*
Women	66.7	50.0	100.0
Men	16.7	0	0
Children	0	0	0
Women and children	16.6	50.0	0
Other	0	0	0
Total	100.0	100.0	100.0

Source: Household survey, August, 1983

fuelwood production. Other men, mainly non-farmers, were observed helping their wives in the collection of fuelwood due to the heavy demands of production. In the fishing village, where fuelwood production is most problematic, 85per cent of the sample who use fuelwood for commercial enterprises have to purchase their fuel requirements.

ECOLOGY AND ACCESS TO FUEL

Ecology sets certain natural parameters for energy supply. The production pattern varies from one ecological zone to another and from one season to the next. It determines to a large extent the labour input of production and the quality and quantity of fuelwood produced at any particular time.

The forest village has the most favourable environment for fuelwood production. The region contains many economic and exploitable wood species, such as cocoa (*theobroma cacao*), opesese (*Funtumia africana*), paya (*Persea gratissima*) and the oil palm (*Elaeis guineensis*) which provide good sources of fuelwood. Indeed, this environment contains so many species of woody plants that women have a choice of both quantity and quality. The most important species used for fuel is opapea (*Phyllanthus discoidens*) which provides a highly combustible wood. Kotobre (*Cassia siamea*), otanuro (*Trichillia hendelotii*) and akakapenpen (*Rouvolfia vomitoria*) are also used frequently.

In the savanna woodland mosaic trees are so scarce that it is difficult to procure fuelwood. Most of the farmland has been almost completely cleared of trees. In addition, expanding population and urban pressures have increased the demand for fuelwood causing deterioration in the original vegetation. Fuelwood production is, therefore, a difficult task that depends on the utilisation of fallow areas.

However, the most important and common tree in this ecological area is the drought-resistant neem (*Azadirachta indica*), which produces excellent fuel. In the savanna villages, 90 per cent of women claimed that it is the most important source of energy. The second most important wood identified is haatso (*Fagara xanthoxyloides*) used by 60 per cent of households. In the fishing village, important resources are yooyitso (*Dialum guineense*) and odanta (*Nesogordonia papaverifera*) both of which produce good charcoal. This is followed by nokotso (*Diospyros mespiliformis*) and onyina (*Ceiba pentandra*) in the savanna and fishing villages, respectively. The latter is not a particularly good fuelwood but is used in time of scarcity.

In the interior savanna where trees have been over exploited the choice of wood species is limited so that taatso (*Millettia thonningii*) which used to be a popular source of fuelwood, is no longer available. Similarly, nokotso and haatso are becoming more and more scarce. It appears that indigenous wood species which are suitable for fuel and are equally useful economically are giving way to the exotic neem which is of less economic value. The growing scarcity of fuelwood has increased both the time invested in production and the cost of purchasing fuel. Even more importantly, it has led to indiscriminate cutting of trees and to soil erosion. For women the destruction of the best fuelwood species has made cooking more difficult.

In both the urban and rural areas women depend almost entirely on rural energy for both income-generating and domestic activities. The evidence so far indicates that due to changing ecological conditions, it is becoming increasingly difficult to obtain fuel. Although the situation is not quite so acute in the forested area, in the interior woodland mosaic zone, there is already a serious shortage of fuelwood. This supports the FAO (1986) prediction that by the year 2000, it is possible that the savanna and semi-arid areas of the country will experience acute fuelwood deficits.

CONCLUSION

Declining rural energy supplies are a challenge to women, particularly in the savanna agro-ecological zone. As a result, the government has become increasingly more concerned with environmental issues. Short-, medium- and long-term strategies are needed to improve and bring about positive changes for women and in the environment (Ardayfio-Schandorf 1987). Existing fuelwood resources need to be maintained and improved while fuel conservation should be incorporated into improved stove programmes in the short term. On a long-term basis, rural development policies should be formulated in order to take account of energy issues. New and renewable sources of energy must be developed while agricultural and agro-forestry strategies may be adopted as an integral part of rural development.

A solution to the problem is not easy, but national governments have got a major role to play with the support of the international community. Local communities and research and development institutions need to collaborate with governments to search out and provide, at affordable prices, alternative energy sources which do not destroy the environment. Such sources should be suitable for the energy needs of women without increasing the need for female labour and time, so as to promote viable development which is both socially and economically sound and environmentally sustainable.

3

WOMEN'S ROLE AND PARTICIPATION IN FARM AND COMMUNITY TREE-GROWING ACTIVITIES IN KIAMBU DISTRICT, KENYA

B. Hyma and Perez Nyamwange

INTRODUCTION

In many developing countries, rural women are primarily responsible for household and farming activities as well as the management of local resources (Hoskins 1979; Williams 1982; FAO 1983, 1985a, 1985b; Fortmann 1986; Dankelman and Davidson 1988). Therefore, women's traditional tasks, experiences, knowledge and concerns about local problems are essential in solving emerging environmental problems such as deforestation, soil erosion and the scarcity of food, fodder, fuel, wood and water. Forestry and forestry-related products are primarily used for fuel, fodder, food, medicine and purposes such as conservation and income generation. Evidence has shown that women have traditionally combined all the above uses, thus accumulating knowledge and values, as well as behaviour patterns that form an integrated approach in their relationship with the environment.

This chapter examines the role and participation of women in tree growing activities in Kiambu district, Kenya. The emphasis on rural women is based on the major role they play in tree planting through their participation in community tree-nursery activities. In this district, local people have adopted certain strategies of managing their ecological habitats, particularly with respect to addressing such problems as soil erosion and deforestation. In essence, the study focuses on participatory processes and on the institutional framework needed to encourage and support its success.

Recently, community forestry has been a major focus of development programmes intended to prevent ecological degradation and to meet survival needs (FAO 1985b). A number of resource management projects have begun to focus on women in the rural areas in order to assist them to use their expertise effectively in local resources management (Hoskins 1979; Williams 1982; Fortmann 1986). Over the past two decades, shortages of food and fuel, rural poverty, rapid population growth, environmental degradation and depletion of forest resources have highlighted the role and significance of effective forest management and the significance of forest products. This is a critical situation, given the overall political economy of production, which is based mainly on cash cropping and equity issues, such as access to and distribution of resources. This paper reaffirms recent views that rural women, despite their long, hard work schedules and their lack of formal education or training, can collectively use the traditional knowledge and awareness gained from daily-living experiences in effectively managing their local environment. They can initiate, plan, design and support programmes to restore or conserve their forestry ecosystems. Given institutional support and representation, women can be a positive and leading force in environmental initiatives and sustainable development programmes.

A major thrust of this discussion is that women's values and behaviour reflect their adaptation to changing environmental conditions. The nature of these adaptations is also determined by the structural constraints in the social relationships and in the institutional context (such as land tenure and security, customs, taboos, beliefs, etc.). Furthermore, women's behavioural strategies are strengthened by the values placed upon sharing, nurturing, reciprocity, co-operation and empathy and by their sensitivity to interpersonal relations and the family (Bernard 1981). The interconnection of female ecological, 'organic' and survival values is evident in the nature of activities undertaken by women in tree growing. The type of tree species preferred, the social organisations engaged in tree-growing activities, the interconnection between motivation and participation, as well as the strategies used are issues that will be discussed on the basis of group interviews with five different women's groups in the district of Kiambu. In general, Kenyan women, particularly Kikuyu women, have been known to be at the forefront of forest management for many years.

THE FOCUS AND STRATEGIES USED IN WOMEN'S TREE-GROWING ACTIVITIES

The issues that community-oriented forestry seeks to address include ensuring a regular supply of energy, maintaining the people's traditional ecological base in order to sustain a flow of material resources and encouraging development based on equity through a sociocultural and economic framework that suits the participants involved. According to Williams (1982), the recent change in the focus of forestry programmes and their related activities from men to women is due to a shift in the development paradigm, which emphasises equity. It is a well established fact that the pattern and process of government distributed opportunities have favoured men in such diverse policies as agriculture, educational training, and reform, extension and other subsidies. Such clear policy preferences for men over women (both in colonial and contemporary times) have led to male advantages in wage employment and capital accumulation (Staudt 1985). However, in Africa, for example, the economic and organisational base continues to build its strength through the spontaneous growth and dominance of women's autonomous self-help associations in private spheres (Staudt 1985; March and Taqqu 1986).

In short, the focus on women involved in tree-planting schemes is justified by the argument: 'since work is socially organised so that women weed and water plants, then there is a need to incorporate them into these tree-growing activities and forestry development programmes' (Hoskins 1979). Additionally, the fact that women know more about the local environment and the social and ecological 'trade-offs' of the various tree species is also an advantage. It has been observed that men prefer timber production for construction and sale, while women have a preference for tree species that offer food and fodder (Hoskins 1979).

In order to implement tree-growing and management activities, various approaches and project designs have been used including participatory projects and collective projects as well as welfare and equity projects (FAO 1983). In planning, designing and programming suitable projects and in ensuring that participants are the beneficiaries, international agencies, government departments and the NGOs concerned have begun to consider viable forms of co-operation, leadership and incentives among women (Dankelman and Davidson 1988). Some of them include indigenous forms of co-operation such as kinship, marital or age groups, as well as women's co-operatives and self-help groups.

The reasons for the above developmental strategies are three-fold. Firstly, forestry programmes intended for local communities are area-specific. It is necessary, therefore, to consider the needs, priorities, perception and knowledge of women in order to maximise the effectiveness of the aims and objectives of a given project (Hoskins 1979; Bronkensha and Alfonso 1984). Secondly, the recognition of the need for an integrated approach involving both collective and individual strategies has prompted development agencies to consider how they can involve both men and women in their forestry projects. Such an involvement has been witnessed in several countries in Africa, Southeast Asia and Latin America (Fortmann 1986). In most cases, the individual and collective strategies are enforced in order to increase women's access to control of forestry resources. (Hoskins 1979; Williams

1982). On a different level, these strategies also reinforce the beliefs and values related to trees and tree products by the local inhabitants. Other reasons include the need to maximise participation of all the social actors as well as to 'utilise the full range of behavioural strategies' such as social and cultural meaning, obligations, rituals, ceremonies etc. Thirdly, community contact groups such as women's co-operatives and self-help groups have provided a ready indigenous organisational and socio-cultural framework within which to operate and implement the intended projects. However, the effectiveness of these groups largely depends on the method of project approach used (Williams 1982; Wanyande 1987). Co-operatives and self-help groups are being used in several developing countries as viable institutions for community participation. They operate successfully under the concept of a 'participatory approach to development' largely because they are decentralised, small, and community based. Initially formed for the purpose of working collectively and for social networking, they now undertake development activities to improve people's socioeconomic situation and the environment (Pala 1976; Wanyande 1987).

Under the concept of *harambee*, literally meaning 'let's pull together', many voluntary groups have emerged since Kenyan independence in 1963, thus providing an organisational framework that facilitates participation. Several of them are women's groups which are largely informal. There are hundreds of women's groups in every province, involved in a variety of developmental activities directed specifically to the perceived needs of the women, the community and the country. Some of the objectives of those groups concerned with the utilisation of natural resources are to increase the supply of fuelwood and timber for construction; to improve their welfare and living conditions and those of their families, and thus become self-reliant, and to improve the quality of their environment and country for future generations. It is this third objective that makes these groups useful in the implementation of various forestry development projects and programmes in rural areas of Kenya, with villages forming the primary units of operation.

FRAMEWORK FOR COMMUNITY FORESTRY PROJECTS: THE EXAMPLE OF THE GREEN BELT MOVEMENT (GBM)

The GBM is a grassroots environmental movement which has set itself a long-term agenda with multiple objectives (Mathaai 1988): to avert the deforestation process through planting trees; to promote indigenous trees and shrubs in order to prevent their extinction; to promote the cultivation of 'multipurpose' trees; to increase public awareness of environmental issues, particularly in relation to population pressure, poverty, migration, food and fuel; to promote socioeconomic development; to create a positive image of women; to make tree planting an income-generating activity for women and to help the rural poor and handicapped.

The GBM's longevity offers a critical insight into the fundamental strengths and weaknesses common to most groups involved in community tree growing and indigenous-based conservation activities. Started in 1977 by the National Council of Women, the GBM was initially concerned with raising awareness about the environment and preventing deforestation. Under the leadership of Professor W. Mathaai, the GBM focused on campaigns and local participation of women in the establishment of 'greenbelt communities' and in the operation of small tree nurseries. It sees women as a powerful leading force for conservation. Though men participate, women are the principal activists and provide the leadership of the organisation and the communities. By 1982, GBM had 500 nurseries each producing between 2,000 and 10,000 seedlings per year, and 239 'greenbelts' had been established (Vollers 1984; Dankelman and Davidson 1988).

An additional objective of the GBM is to promote co-operation with government authorities and the public forestry administration, including forest officials and extension personnel, District Development Committees (DDCs), NGOs and other conservationists. In this way, the GBM encourages, organises, sponsors, coordinates, supervises and trains members of several women's groups in tree growing activities. This central role is very important since the efforts of other self-help groups are identified with GBM. Participation by community members in managing nurseries and growing tree seedlings and tree planting is taken as part of the farming system that is linked to other agricultural activities. A major role of GBM to other community groups is the provision of technical assistance at national and international levels (Figure 3.1).

The GBM has been successful in decentralising forestry activities for several reasons. Firstly, the group is based on a Western concept of institutional arrangements. It has a set of objectives, a chairperson, a secretary and a committee that are chosen by its members. Therefore, it is contributing to the institutionalisation of community forestry and implementing activities with the community. Responsibilities and activities are shared equally among the various members. It acts as the central body in seeking sponsorship, in allocating funds, activities and responsibilities to other branch members. Secondly, the group is directed by women leaders who are either well educated or the wives of influential people with connections in the mainstream political spheres (or both). The GBM works closely with the forest department in publicising tree

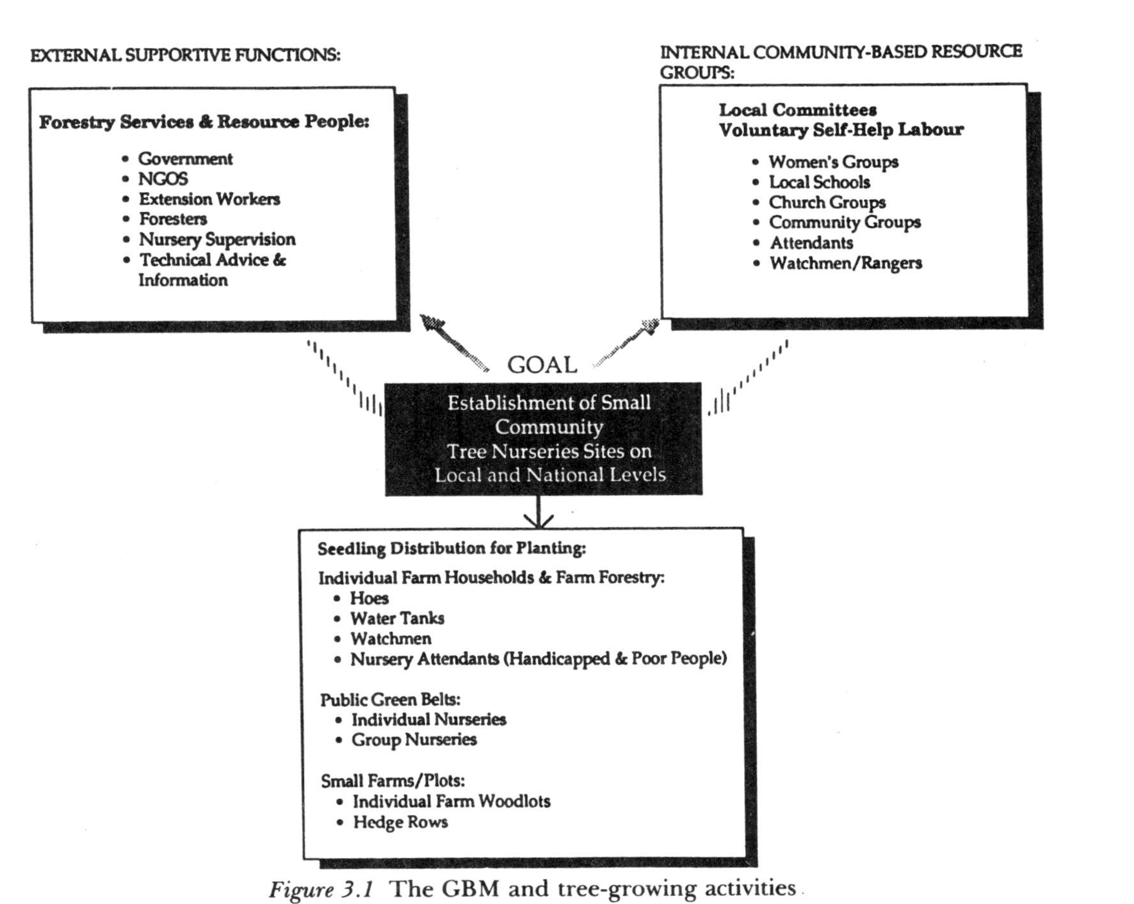

Figure 3.1 The GBM and tree-growing activities
Source: P. Nyamwange and B. Hyma

planting and in the purchase of seedlings. The leaders are able to use their position to get assistance in terms of funding and technical and professional expertise through extension services. They have direct access to the resource people and external agencies, unlike many scattered and unintegrated women's groups in the rural areas. Local organisations can play a valuable role in implementing tree-planting programmes. They motivate people and bridge the gap between the people and the forest department.

However, these very facts point out the principal advantages of national-level institutional arrangements and how they retard the performance of rural-based informal women's groups. Based on the responses from our survey, some of the criticisms against GBM include bias in their assistance including overly high qualification requirements (for example, to be enrolled as a member, interested groups must pay a monetary contribution to the movement), manipulation of the GBM activities by politicians and the general authoritarian attitude of the GBM's executive members towards smaller groups. For example, if a group does not have a structured organisational base for its activities, it does not qualify for assistance.

CASE STUDY: KIAMBU DISTRICT

Kiambu, with an area of 2,448 km^2, is one of the six districts which make up the Central Province of Kenya. The district is situated in the south-western part of the Province; Nairobi and Kajiado districts form the southern border. Kiambu's location in a high-rainfall zone is an advantage, for it enjoys both forest and water resources as well as high potential arable land. Its proximity to the capital city, Nairobi, is advantageous for it provides a ready market for the agricultural products of the district.

Kiambu is the second most populous district in Kenya, with a total population of 686,290 (1979), which is projected to rise to 2.1 million by 2010. The current population density is 340 persons per km^2 and is estimated to reach 854 persons per km^2 by 2010. This projection has serious implications for the district's natural resources. According to the 1979 census, the district had a population of 344,366 males and 341,924 females (Table 3.1). Gatundu division has the highest population (152,371) and females outnumber males at a ratio of 108:100. Gatundu is also the least developed part of the district, experiencing severe soil erosion due to the deforestation of the hills.

Table 3.2 illustrates the land categories in the district, which has seven administrative divisions. Most of the land has been described as suitable for various food crops as well as commercial crops.

Table 3.1 Population of Kiambu district, by division

Division	*Males*	*Females*	*Total*	*% women*
Kiambaa	52,228	51,831	104,059	49.8
Gatundu	73,012	79,359	152,371	52.0
Lari	30,412	32,117	62,529	51.4
Limuru	36,213	34,924	71,137	49.0
Thika	51,467	39,486	90,953	43.4
Githunguri	46,816	49,279	96,095	51.3
Kikuyu	54,218	54,928	109,146	50.0
Total	344,366	341,924	686,290	49.8

Source: Central Bureau of Statistics, Census 1979.

Table 3.2 Land categories: Kiambu district

Land category	*Area (sq km)*	*% land*
Government land		
Forest reserve	448	
Govt reserve land	65	
Townships	98	
Alienated land	446	
Unalienated land	10	
Total	1,067	43.6
Freehold land		
Smallholder schemes	20	
Others	291	
Total	311	12.7
Trust land		
Townships and others	58	
Smallholder schemes	1,012	
Total	1,070	43.6
Open water	3	0.1
Total	2,451	100.0

Source: Central Bureau of Statistics 1983

The estimated area under forests in the district is over 44,876 hectares. Of this, 33,000 hectares are under indigenous trees. Additionally, there are various woodlots owned by individual farmers which cover an estimated 450 hectares. According to the Tana and Athi River Development Authority (TARDA), the district is capable of meeting its demand for fuelwood and timber if certain development strategies are adopted. These include: improving existing trees and forests; establishing individual holdings with woodlots; reducing reliance on woodfuel by developing alternative sources of energy; and decreasing the number of fruit trees. Other goals include the provision of fuelwood and house construction material for the community; the prevention of soil erosion; the generation of income and employment; improvement of the village environment and the provision of dry-season fodder for cattle and other livestock through community forestry projects (FAO 1986).

FOREST DEVELOPMENT PROPOSALS

The Ministry of Environment and Natural Resources proposes to increase the capacity of the district's nurseries from 2.6 million to 5 million seedlings by the year 1992. Table 3.3 outlines the current nursery ownership patterns and seedling capacity. According to TARDA, the Ministry is not capable of implementing the additional 2.4 million seedlings. It is recognised that the supportive institutional environment in which community forestry programmes take place is crucial for its success (FAO 1985a). It is therefore suggested that other agencies such as the Green Belt Movement, schools, local chiefs, and other government institutions undertake the distribution of seedlings to individual farmers. In addition to two operational nurseries propagating temperate fruit tree seedlings, both managed by the Prisons department in Kamiti and Kiambu, the Ministry proposes to establish seven additional agro-forestry nurseries with capacity for 100,000 seedlings to produce high-quality apples, pears, avocados and fodder trees and grasses for distribution to farmers at nominal prices.

The introduction of multipurpose tree-growing activities through decentralised community nurseries has several advantages. Firstly, it recognises the different potential and needs of individual farm units. Secondly, although rural people can obtain naturally germinating seeds, nurseries are able to provide a broader range of species, including

Table 3.3 Existing nurseries and seedling capacity: Kiambu

Name of institution	*No. of nurseries*	*Seedling capacity*
Rural forestry		
Extension service	3	900,000 (35%)
District development		
Committee	3	176,000 (7%)
Chiefs	26	600,000 (23%)
Greenbelt movement	17	300,000 (11%)
Individual owners	12	200,000 (8.5%)
High schools	3	6,000 (0.2%)
Primary schools	15	90,000 (2.2%)
Large-scale farms	2	60,000 (2%)
Training and research institutes	3	300,000 (11%)
Prison departments	2	3,900 (0.1%)
Total	86	2,635,900 (100%)

Source: TARDA, 1985 (unpublished)

improved indigenous varieties. Thirdly, seedlings grown under controlled conditions will provide a degree of quality control and selection from strong parent stock. Finally, the nurseries can be a focal point for technical advice and extension service to both men and women farmers.

WOMEN'S CONCERN FOR ENVIRONMENTAL ISSUES IN KIAMBU DISTRICT

Women in forestry activities

Women provide 60–70 per cent of farm labour in Kiambu district. They are responsible for both food and cash crop production. Indeed, the quality of their lives as well as their environment depends entirely on how the women organise their day-to-day activities. The quality and quantity of farm production depends on access to land, agricultural inputs, labour, credit, extension services and land use patterns.

The district, with its hilly topography and problems of population pressure, faces erosion and conservation concerns that place severe pressure on both land and forest resources. In response to this situation, the women have spontaneously adopted an integrated approach through multiple land use, such as multiple cropping, inter-cropping, bench terracing and cultivating both annual and perennial crops, all of which can be considered as part of the traditional farming methods.

In general, the concern over the quality of their environment, coupled with shortages of fuelwood and fodder during the dry season, has made the women sensitive and cautious in dealing with land and forestry resources. Women are motivated to grow trees not only for fuel but also for fodder, fruit, as windbreaks, for fencing and for shade and construction materials. This fact supports the notion that fuelwood shortages cannot be isolated from other problems facing the women in the rural areas (FAO 1986).

The role of women's self help groups in tree-growing activities

There are approximately 2,000 women's groups in Kiambu district with a total membership of 70,000. These multipurpose groups include clan and church groups, as well as local community groups that are engaged in small-scale and large-scale development activities. Such activities include income-generating work such as basketry, pottery, bee keeping and brick making as well as conservation activities such as tree growing and nursery management.

The activities and efforts of the women farmers are sustained by the assurance that the benefits of their labours are for the collective good of themselves. One surveyed group stated that they use a participatory management style whereby all its members have a shared interest in managing the tree resources. Some active members of these groups have individually taken the initiative to organise their own tree nurseries as well as growing trees on their farm lot. This practice is supported further by the regular visits by other members and through information sharing.

WOMEN AND TREE-GROWING

Five women's groups in Kiambu district were interviewed (Table 3.4). Women's involvement in forestry resource management, as reinforced by GBM in Kiambu, supports the idea that women have managed and will continue to manage and develop farm woodlots, orchards, and multipurpose tree-growing activities and projects. This fact is evident from some of the activities undertaken by the women surveyed (Table 3.5). Firstly, the projects reveal an integration of traditional practices and modern concepts of forestry management. For example, the women have a 'collective orientation' towards their tree growing activities even though the trees are privately owned. Coupled with this, the traditional practice of growing trees that have multipurpose products have largely aided their activities. Secondly, the women's knowledge of tree species and suitable ecological conditions is an advantage that they have used in selecting the area and type of trees they grow, even with food and cash crops. Furthermore, these activities have been extended and used to improvise the women's low incomes through the sale of forestry products. This objective has been realised either individually or through the various women's self-help groups in the district.

Table 3.4 Women's groups interviewed for this study

Name of group	*Location/ division*	*Total membership*	*Number interviewed*
Kanyariri Mother's Union	Wangige, Kikuyu	800	100
Mwimuto Maendeleo Ya Wanawake	Ngoroko, Gatundu	350	50
Kireita Catholic Women's Assoc.	Kireita, Lari	250	180
Ikinu Home Industry Women's Group	Ikinu, Githunguri	650	120
Nderi Women's Group	Nderi, Kikuyu	100	30
Total		2,150	480

Source: Author's fieldwork

Table 3.5 Tree growing and other activities of women's groups

Name of group	*No. of nurseries*	*Tree seedlings*	*Other activities*
Kanyariri Mother's Union	5	10,000	Selling seedlings, homecrafts, collective labour
Mwimuto Maendeleo Ya Wanawake	1	200	Charcoal production literacy classes, homecrafts
Kireita Catholic Women's Assoc.	4	2,000	Collective labour, cookery, counselling homecraft, church related activities
Ikinu Home Industry Women's Group	3	7,000	Pottery, stovemaking, selling tree seedlings, land business
Nderi Women's Group	5	500	Fund raising, collective labour

Source: Author's fieldwork

The groups interviewed expressed several reasons for their participation in tree-growing activities. Firstly, they are aware of the benefits of tree growing activities, such as soil and water conservation. Secondly, they have the desire to better the quality of the environment for future generations. For example, one group expressed the need to ensure that their children have a regular supply of water and fuel. Thirdly, they recognise the benefit of multipurpose tree activities in generating income, improving soil quality and meeting a wide range of household needs (such as fibre, fuelwood, shelter and medicine). Another reason that motivates these groups in their activities is the desire to grow and preserve indigenous tree species. Finally, they felt that modern methods of tree growing are not suitable for their needs and that their active participation in these activities is to learn something new while preserving the old and building on existing strengths and capabilities.

Participants are stimulated and motivated by other factors. The women know that they need to ensure a regular supply of fuelwood for domestic use. The Ikinu Home Industry Women's group has encouraged its members to grow trees in the plots where they cultivate food crops. Alternatively, the group has encouraged the use of energy-saving stoves by its members. Increasing demand for wood products, fruits, nuts, edible leaves and animal fodder, tanneries, dykes, barks, fibre, medicinal plants, gum and oil have also been a significant motive for tree planting. Secondly, the recognition and assistance of their conservation activities by the government have encouraged women further. This encouragement is evident from the provision of inputs such as free tree seedlings and in the increase of extension workers and services tailored specifically to women and their interests. The Nderi Women's Group, for example, is motivated by the regular visits to the group's tree nurseries by the local forester as well as by the local community development agent.

CONSTRAINTS FACED BY THE WOMEN'S GROUPS

This broad survey of selected women's groups has concentrated on attitudes towards deforestation and its consequences, and current tree-planting activities. Special attention was given to reasons advanced for the limited amount of planting. All five groups interviewed identified similar constraints to their tree growing activities: lack of sufficient inputs; infestations; shortage of water, manure, fencing and containers; a continuous lack of organisational and management skills; lack of proper training; scarcity of land; time constraints; lack of capital to enhance activities; irregular extension services and non-recognition of their activities by the government. Additionally, it was observed that heterogeneity both within each group and between the groups (with respect to age, education, status, motivation) proved somewhat problematic. For example, groups with members who are 'better off' are favoured in terms of funding and extension services. Also, the heterogeneous nature of these groups implies disunity and confusion among its members, which may be reflected in their activities. The dismissal of the local chairperson of Maendeleo Yo Wanawake in 1986, due to mismanagement of funds and favouritism, is a case in point. It is not surprising therefore, that most of these women's groups are treated with caution (as far as donor agencies are concerned) and suspicion by their male counterparts (Tostensen and Scott 1987).

Women's participation is constrained in many ways. Institutional and organisational support for education and training of women in modern community forestry activities is far from established. Other factors such as the lack of access to and participation of women in the local village council and DDCs, limit the effectiveness and efficiency of the operations. Moreover, only a few professional foresters are women and the range of training available to rural women is very limited. Community tree nurseries also lack sufficient funds for staff help, seed production, and land for expansion. Their potential to serve as mini-extension centres is far from being achieved. They remain on a scattered and small-scale basis while the rate of deforestation and energy demands is accelerating. Proper planting and care are still not appreciated on a large scale. Nurseries cannot be run on a voluntary basis; incentives and innovative programmes are needed and motivation in the form of financial gain and paid staff is important.

DISCUSSION

An analysis of the case study suggests that there is a critical need for forestry activities throughout Kenya to adapt to changing societal needs. This is necessary if forestry issues are of value and concern to the community and remain within its control. It is evident that the traditional behaviour of women regarding forestry resources and regeneration gradually evolves according to their needs, the nature of the ecological base and society at large.

The tree-growing activities of the women in Kiambu district are most encouraging to those whose main concern is to protect the environment. Yet, their slow progress indicates a continued lack of co-ordinated assistance and a lack of women in the forestry service. This inequity can be partially explained by the conditioning and attitudes of men and an underlying effort to maintain the traditional gender division of labour. However, it must be pointed out that this official attitude and bias against an activity such as tree growing, which is not immediately productive in the sense of cash-crop production, is not dictated by gender prejudice alone. The bias is also tied to the class issue and the entire response to the changing environment which has its roots in the colonial trade tradition and the dominance of export-oriented cash-crop production.

It is clear that women have different values, needs, and strategies for action in forestry activities. However, there is a need to change or review bureaucratic forestry institutions and their policies and to increase the number of women forestry professionals and extension agents. These changes are necessary in order to incorporate women's community forestry activities into the main stream of development.

Observation of the Kiambu case study indicates that women's values and behaviour reflect their adaptation to both the state of environmental conditions as well as structural constraints in the sociocultural and institutional context. For example, different women's groups have used other means to overcome time and money constraints and proceed with achieving their aim. This statement also applies to external assistance and the lack of organisational skills.

This paper has argued that agricultural and ecological values overlap thereby emphasising the 'interconnectedness' of agriculture and forestry. The example of women's tree-growing activities in Kiambu district shows how their knowledge, experiences and values in both activities are interrelated and interdependent. Further, women's behavioural strategies (such as forming groups in order to get government assistance) are strengthened by the values placed on 'sharing, nurturing, reciprocity, cooperation, empathy and their sensitivity to interpersonal relations and family' (Bernard 1981: 21). This paper also confirms recent perceptions that rural forestry innovations must be based on examining traditional tree-management practices and indigenous knowledge, of both women and men, before introducing new management strategies. According to FAO (1985b), 'rural people usually maintain or grow trees to provide multiple outputs rather than single products such as fuelwood'. Interviews with five different women's groups reaffirms this belief.

A successful community forestry programme needs to satisfy a range of different measures such as education, training and extension work (FAO 1985a). It is recommended that women's participation and representation in education, training, extension services and decision-making roles be increased or enhanced considerably in community forestry and conservation development programmes. Recent awareness of the central roles that women play in tree utilisation and maintenance points out the need for the inclusion of women in agricultural extension activities and that training for women's groups should be given high priority in research funding and training programmes. Since the quality of their environment almost entirely depends on their own activities, new methods of professional education and training have to be established for existing forest officials and extension personnel; methods that are sensitive to their changing roles. These roles range from traditional executive style administration to supportive inputs and technical advice, training, guidance and extension services for participatory-style management by the community involved in the specific social forestry projects (Shiva 1987; Sanwal 1988).

CONCLUSION

This survey of selected women's groups in Kiambu reaffirms the central role of women in forest utilisation and maintenance. An overall examination reveals that women are among the most experienced rural actors and resource managers. They have proved a positive and potential force in both restoring and conserving the rural environment in Kenya. Hence, the institutionalisation of community forestry must be designed specifically in relation to women's needs and their environmental concerns. Planning programmes need to reflect the existing indigenous knowledge of tree management and conservation practices of both women and men.

Promoting and strengthening household/farm/community tree management depends on the existence of strong local self-help groups (FAO 1985a). Although rural women leaders are emerging within women's organisations, often the reluctance of local administrations (village councils, district development committees) to encourage formal representation of women on the heavily male dominated local councils and committees has prevented the growth of village level capabilities (Staudt 1985).

Although separate membership organisations for women, like the GBM, provide greater autonomy of expression and opportunities for women to develop skills, leadership and self-confidence, they often limit their political representation in decision-making bodies at the district level. Integrated organisations which equally represent men and women may be an ideal approach for creating sustainable rural development (Uphoff 1986).

Women and men have shown preferences for multipurpose and different species of trees. An integrated approach to women's local concerns in agricultural development is more likely to be beneficial in the long run. More research is needed into agro-forestry practices and into possible innovations involving women and men. This requires decentralised research and development programmes for developing communal or farm forestry designs that respect the different objectives of rural people experiencing different circumstances in their own environment.

Seedling production and distribution should be centred on a large number of small nurseries, ideally selected from dispersed farms, villages and households. This could facilitate people in individual villages to develop better appreciation and awareness of trees and tree management as well as raise their income level. It has been shown that an increase in species selection and propagation and villagers' positive response to plant seedlings is in part a function of the village's proximity to nurseries (FAO 1985).

Community forestry projects require a new breed of forester who is trained to deal with both the technical and human dimensions of forest conservation and development. FAO (1985) recommended that strong, clear laws and regulations need to be developed in agreement with village organisations and groups. Overall, many writers have acknowledged that community forestry programmes are among the most difficult projects to design and implement on a large scale. However, empirical evidence is emerging that indicates that large-scale tree-planting programmes, managed by village councils, cooperatives and community groups can be successful.

4

AGRICULTURAL PRODUCTION AND WOMEN'S TIME BUDGETS IN UGANDA

Victoria M. Mwaka

INTRODUCTION

This chapter examines rural women's problems of time budgeting and agricultural production in eastern Uganda. It focuses on the effects of drudgery and time limitations on women's contribution to the agricultural sector and outlines possible remedies for the plight of these women.

AGRICULTURE AND ENVIRONMENT

Uganda is a small country in the heart of Africa with 16.5 million (1989) people, of whom 53 per cent are female, on 241,000 square kilometres. The country is well endowed in terms of agricultural resources and 84 per cent of the population farm the land. Much of the country receives 1500 mm of rain and even the driest areas receive 500 to 750 mm per annum with only 6 per cent of land being classified as semi-arid. Permanent cropping of both cash and food crops is possible in most parts and 34 per cent of the nation is under permanent cultivation. Agriculture contributes half the gross domestic product and coffee makes up 97 per cent of the exports by value. The strength of agriculture lies in its organisation based on smallholdings with an average farm size of 2.5 hectares from which 95 per cent of the nation's food requirements come.

It is evident from Table 4.1 that in recent years production of food crops has fluctuated and in some cases has actually declined. Agricultural output fell by 30 per cent over the period 1970 to 1985 and per capita GDP by 43 per cent. This fall was not due to natural catastrophes like droughts or floods but was the result of the adverse human factors of war and the world economic recession. The agricultural economy has been operating within a deteriorating national framework but one of the

Table 4.1 Production of principal food crops in Uganda, 1973–84

			'000 tons		
Crop	*1973*	*1976*	*1979*	*1982*	*1984*
Maize	419	674	453	400	416
Finger millet	643	576	481	528	549
Sorghum	389	390	316	400	410
Rice	–	12	5	9	10
Plantains	8,126	8,137	6,090	6,600	4,950
Sweet potatoes	1,232	2,002	1,272	1,600	1,857
Cassava	2,132	2,838	2,110	3,300	3,264
Pulses	260	420	203	361	502
Oil seeds	243	210	132	124	179
Sugar (raw)	69	18	5	–	–

Source: Ministry of Planning and Economic Development, 1987

underlying causes of the decline of the agricultural sector has been neglect of the main producers, rural women.

WOMEN AND AGRICULTURAL PRODUCTION IN UGANDA

In Uganda women are involved in the production of subsistence and cash crops. Basic household requirements should be the responsibility of men but in practice in the rural areas some women provide even the basic needs of the family. Women grow the major cash crops such as coffee and cotton, cereals such as maize and millet, basic subsistence crops like cassava and plantains and also rear cattle, pigs, goats and chickens.

In many African societies, household relationships are characterised by inequality in the distribution of work, land, income, consumption and contribution to productivity, based on gender and age. Household inequalities are reinforced in the juridical, political and traditional/cultural realms where family members are considered dependants of the head of the household who is taken as spokesman and legal representative of the household. The man will always dictate to the woman what to do but he does not care how she finds time to do the work. Sometimes her work is too time consuming but if the woman fails to complete her tasks through lack of time she may be reprimanded with insults and severe beating. Such gender inequality is reflected or manifested in many problems that affect agricultural productivity in Uganda. These problems include women's heavy burden of drudgery, their land shortage, their lack of institutional support in the form of credit, extension services and appropriate technology, inadequate infrastructure in the form of markets, transport and health care and finally, in general, female apathy.

Women's burden of work and time constraints are among the major causes of the deterioration of agriculture in Uganda. Earlier concentrations on the production of cash crops for export led to stagnation in food crop output for internal markets and home consumption because it increased the burden on already over-worked rural women. Time-consuming and heavy work like bush clearance and land preparation, planting, weeding and harvesting of cash crops is usually done by women. In addition, women are expected to produce enough food for household subsistence plus a surplus for sale to earn some extra income with which to purchase other basic necessities such as soap, salt, children's clothing and medication. Women also have to undertake traditional household tasks such as fetching water, collecting firewood, and taking care of home sanitation. These tasks are a heavy burden in a country where the fertility rate is 6.9 (1985–90); 49 per cent of the population was under 15 years of age in 1985 and households of ten people are not uncommon.

This disproportionate load on the backs of rural women is rendered even more problematic by the very limited support given in terms of technology designed to lighten household and farm work. Table 4.2 shows what appears to be an equal division of labour between wife and husband in rice growing. However, in the case of men, farm work is only done during the morning hours with afternoons and evenings spent in relaxation with local refreshments and socialisation with village mates. For women, it is a different story as Table 4.3 illustrates. There is no time for evening rest or beer drinking and socialisation. Mornings and afternoons are devoted to farm work and the mandatory household chores also have to be fitted into the women's day. It is evident that women's problems of time budgeting in Uganda are not very different from those of rural women in other developing countries (Momsen 1991). Many such women spend more than 15 hours each day in specific tasks which are often combined with child care which is a constant responsibility throughout the day and night.

Table 4.2 Divisions of labour in rice growing, eastern Uganda

Activity	*Participant*
Land/bush clearing	Husband and wife
Planting	Husband and wife
Weeding	Husband and wife
Bird scaring	Children and wife
Clearing canals	Husband and wife
Harvesting	Husband, wife, hired labour and group work
Threshing	Wife (machine aided)
Winnowing	Wife (machine aided)
Selling	Husband

Source: Fieldwork, 1988

Table 4.3 Workday of a rural woman, eastern Uganda

Type of work	*Duration*	*Time spent*
Preparing breakfast for household	6.00–7.00	1 hour
Working on the farm	7.00–11.00	4 hours
Sundrying harvested crops	11.00–11.30	30 minutes
Fetching water/firewood	12.00–12.30	30 minutes
Taking goats to graze/feeding pigs	12.30–14.00	90 minutes
Preparing and serving lunch	14.00–15.00	1 hour
Working on the farm	15.00–18.00	3 hours
Fetching water/firewood	18.00–18.30	30 minutes
Preparing dinner and bathing children	18.30–20.00	90 minutes
Boiling water for bathing for adults	20.00–20.30	30 minutes
Serving evening meal	20.30–21.30	1 hour
Tidying the house	21.30–22.00	30 minutes
Total working time		*15.5 hours*

Source: Fieldwork, 1988

Food production in Uganda usually takes second place as far as crop land allocation is concerned. The largest proportion and most fertile part of the land owned by the family is commonly devoted to cash crops by order of the household head, the man in most cases. The hoe is still the main cultivation tool. In areas where ox ploughing is practised, the death of oxen and the continuing shortage of spare parts for ploughs are increasingly forcing many farmers to revert to hand cultivation. Despite an emphasis on tool distribution by donor agencies during the last decade, lack of tools is one of the greatest problems felt by women farmers. Tractors and other modern implements are expensive to purchase and maintain and their acquisition can only be contemplated by a very small minority of the most well-to-do farmers. Therefore, mechanisation is rarely extended to women despite the fact that rural women in Uganda are responsible for 80 per cent of food production. Hand tools make work slow and consume much human energy and time. The amount of work done with hand tools depends on an individual's stamina, carefulness and dedication. Hired labourers are scarce and expensive in rural areas as each household concentrates on their own farms.

Higher-yielding seeds, new tools and machines, fertilisers, credit facilities and training in improved farming techniques have been largely unavailable to Ugandan women. Agricultural extension services only reach the so-called progressive farmers, who are men in the vast majority of cases, and advice is generally given only for export crops. Extension services are rarely provided for food-crop farmers, most of whom are women.

Infrastructural facilities constraining women in their agricultural and household chores include transport, communications, and food processing and marketing facilities. Men have their bicycles and can afford taxis, buses and lorry transport. However, most rural women have to walk long distances in order to accomplish necessary journeys for household reproductive activities such as taking children to hospital and collecting water and firewood. Efforts to increase agricultural production are frustrated by lack of easy access to rewarding markets.

With regard to crop processing, men take produce intended for sale to local mills in the rural areas. For home consumption, women use hand tools to grind grain, shell peanuts and mill rice during the daily preparation of family meals. This is very time consuming and tiresome, and eats into time that could have been spent on other activities more personally or economically rewarding.

Communications and information flows are bottlenecks in rural areas. Most women rarely leave their homes to find out what is happening beyond their compound and neighbours. Time constraints and household chores do not allow them to make 'unnecessary' journeys. Necessary journeys are those for collecting water and firewood, taking children to hospital and going to burial ceremonies. Newspapers do not reach most rural areas and when they do it is predominantly men who have the time and education to read them. Batteries for radios are often in short supply and again women do not have time to listen to broadcasts. Many rural women in Uganda are cut off from society other than in their own homes and villages. This isolation hinders their ability to adopt new farming techniques, let alone acquire new ideas for self-improvement.

The overall end-product of all these constraints is that insufficient food is being produced to sustain families throughout the year. The drudgery of women's lives, the fragmentation of their time and the competing demands of cash crops for labour reduce their ability to devote time to food production. Time pressures may lead to changes in land use and family diets. Bradley (1991: 71) has shown that in neighbouring Kenya the time and energy demands of millet, whose cultivation, harvesting and processing takes twice the labour of a similar area of maize, have led to declining production of this crop, especially among younger women who are learning to value their time more highly. Table 4.1 shows a 15 per cent fall in production of millet in Uganda between 1973 and 1984.

REMEDIES FOR THE PLIGHT OF RURAL WOMEN

The government of Uganda recognises the importance of the role of women in agriculture and in development in general. A Ministry of State for Women in Development, based in the President's Office, and headed by a woman, has been set up recently. The Minister for Agriculture and Forestry is a woman and the Deputy Minister of Youth, Culture and

Sports is also a woman. The major objective of the Ministry for Women in Development is to integrate women into the development of the nation. The Ministry is however in need of relevant data on women's activities as a basis for policy formulation, ultimate decision making, programme and project identification and implementation for the improvement of conditions for women, especially in the agricultural sector.

The Uganda Commercial Bank has instituted a Rural Women's Credit Programme through its chain of banking outlets in the rural areas. The scheme is intended to help women to acquire credit without collateral. Such credit would be used to open small businesses in rural areas as well as assisting in large-scale agricultural production. Despite the availability of this facility, women have not taken up the programme with vigour because of their inferior educational standards, and lack of self-confidence and general entrepreneurial skills. A major constraint is that few women have their own land on which to locate businesses or carry on large-scale farming. However, this may change, and the future for Ugandan women may begin to look brighter.

It is widely advocated that poor rural women themselves must assume the ultimate responsibility if their conditions are to improve. They need an operational base to act as a link between themselves and bureaucratic agencies. It would help to make women in rural areas aware of available agricultural resources. Organisations or co-operatives are needed for specific purposes to provide joint or common services, credit, marketing, and training as well as facilities for child care and family planning with a view to reducing women's workload while increasing their income-earning opportunities and productivity in the agricultural sector.

5

THE IMPACT OF LABOUR-SAVING DEVICES ON THE LIVES OF RURAL AFRICAN WOMEN

Grain mills in The Gambia

Hazel Barrett and Angela Browne

INTRODUCTION

During the 1970s rural development was highlighted by the World Bank as the focus of aid policies and development strategies. This was a major departure from previous policies which stressed that the benefits of urban and industrial investment would trickle down to rural areas. At this time development was seen in purely economic terms with the World Bank in 1975 defining rural development as 'the modernisation and monetisation of rural society, and . . . its transition from traditional isolation to integration with the national economy' (World Bank 1975: 3). In most developing countries this was interpreted as the need to introduce new agricultural technologies and innovations, most notably those associated with the green revolution (Todaro 1977). However, the failure of these technologies to meet the social and economic needs of rural populations, particularly women, resulted in a reassessment of the goals of rural development, with a distinct move away from purely economic considerations. By the 1980s, the achievement of a perceptible and sustained increase in the standard of living of the low-income population living in rural areas had become the accepted goal of rural development (Mabogunje 1980: 94). It had also become clear that if rural development was to succeed, then women had to be given a higher priority: as the Decade for Women documentation stated: 'the role of women in development . . . is fundamental to the development of all societies' (UN 1985: paragraph 12).

Improved technologies were believed to hold out considerable benefits for rural communities and were seen by governments and aid agencies alike as politically neutral and acceptable solutions to rural poverty. Labour-saving devices, it was argued, would relieve rural folk of many back-breaking and under-productive chores on the farm and in the home and thus increase labour productivity. The theory was that the time saved could be diverted into income-generating activities, better child care and a general increase in the well-being of the whole family, and thus result in rural development (Carr 1985). In many areas of the Third World women are the major producers of food crops and perform most domestic tasks (Rogers 1980; Adeyokunnu 1984; Momsen and Townsend 1987; Ivan-Smith *et al.* 1988). These are often performed using traditional methods with few modern aids and are thus very time and energy consuming. Developmentalists argue that women are so heavily occupied in under-productive tasks that they are unable to participate fully in the development process; as a result rural development in many areas is being slowed down (Sandu and Sandler 1986; Barrett and Browne 1989). This is a situation that can be resolved by the introduction of relatively unsophisticated, socially beneficial technologies such as hand- or foot-operated pumped wells, fuel-efficient cooking stoves and mills to process cereals. This, it is argued, will release women from time-consuming domestic duties and enable them to engage in income-generating projects (Carr 1985). One of the most time-consuming tasks facing women in all Third World societies is the preparation of food, and for this reason, the governments and donor agencies have been keen to introduce technologies that alleviate the drudgery of cereal processing. This chapter will investigate the impact of one labour-saving device, the mechanical cereal mill, on the lives of rural women in The Gambia, focusing on its actual and potential developmental benefits for women and the community in general.

THE GAMBIA

The milling of grain is one of the most basic of crop processing requirements (Intermediate Technology 1985). The traditional method of grinding in The Gambia is by pestle and mortar. This equipment comprises a hollowed-out section of tree trunk and a large pole, up to 2 metres long and weighing about 10 kg. During the grinding process, the pole is rammed repeatedly into the grain in the mortar, causing women to suffer from backache, tiredness and sore hands (Intermediate Technology 1985). Grinding rates are low, estimated at less than 0.5 kg per hour. Families in The Gambia eat cereal two or three times a day, rice and millet or sorghum (locally known as *coos*) being the major staples. *Coos* has a greater nutritional value than rice (UNIFEM 1988) and is eaten once or twice a day, depending on household supplies. However, it takes four times longer to prepare. The introduction of power-driven *coos* mills is therefore an attractive proposition, potentially releasing women from up to 4 hours energy-consuming cereal processing each day – time that could be used for developmental activities or even leisure.

Since 1980 aid agencies have been installing *coos* mills throughout the country. Donors have included Catholic Relief Services (CRS) and ActionAid, both non-governmental organisations, and UNIFEM and the Canadian government, through the Gambia Women's Bureau. By late 1988, there were about 30 mills in operation throughout the country (Figure 5.1). There are several types of machine but all are diesel powered, and usually comprise a hammer mill and huller. The Women's Bureau (UNIFEM) mills are the largest and most expensive, with total capital costs shown in Table 5.1, and with the two machines housed in a two-roomed building each with its own diesel-powered unit. The CRS mills are smaller, and the huller is locally manufactured. In this case, the two machines are housed in a one-room shed, being mounted at either end of a single diesel engine to which each can be linked by a drive belt. The total costs are therefore lower, at about one-third that of the UNIFEM mills.

Rice and *coos* are grown in all parts of The Gambia, but the proportion of cultivated land under these crops varies regionally. From Figure 5.2, it can be seen that Western Division (34 per cent) and Upper River Division (28 per cent) have the highest proportion of land devoted to *coos* production, while Lower River Division (12 per cent) and North Bank Division (15 per cent) have the lowest. However, there appears to be no relationship between the location of the *coos* mills and the proportion of *coos* land, which may partly explain why, in some areas of the country, mills are not economically viable.

Table 5.1 Capital costs of UNIFEM *coos* mills

Item	*Delasis (D)*	£
Hammer mill	9,504	864
Huller	27,184	2,471
Building and concrete plinth	25,600	2,327
Total	D62,288	£5,662

Source: Women's Bureau (unpublished) 1988 (£1=D11, mid 1988)

Impact of *coos* mills

For the present research, all the sponsoring and funding agencies involved in supplying *coos* mills were questioned and an inventory of such machines compiled (Figure 5.1). One village, Basori, was studied in detail by means of a questionnaire survey of all households in the village, during the rainy season of 1988. Five major themes are addressed by this research: access to the mill; time and energy saved and use of

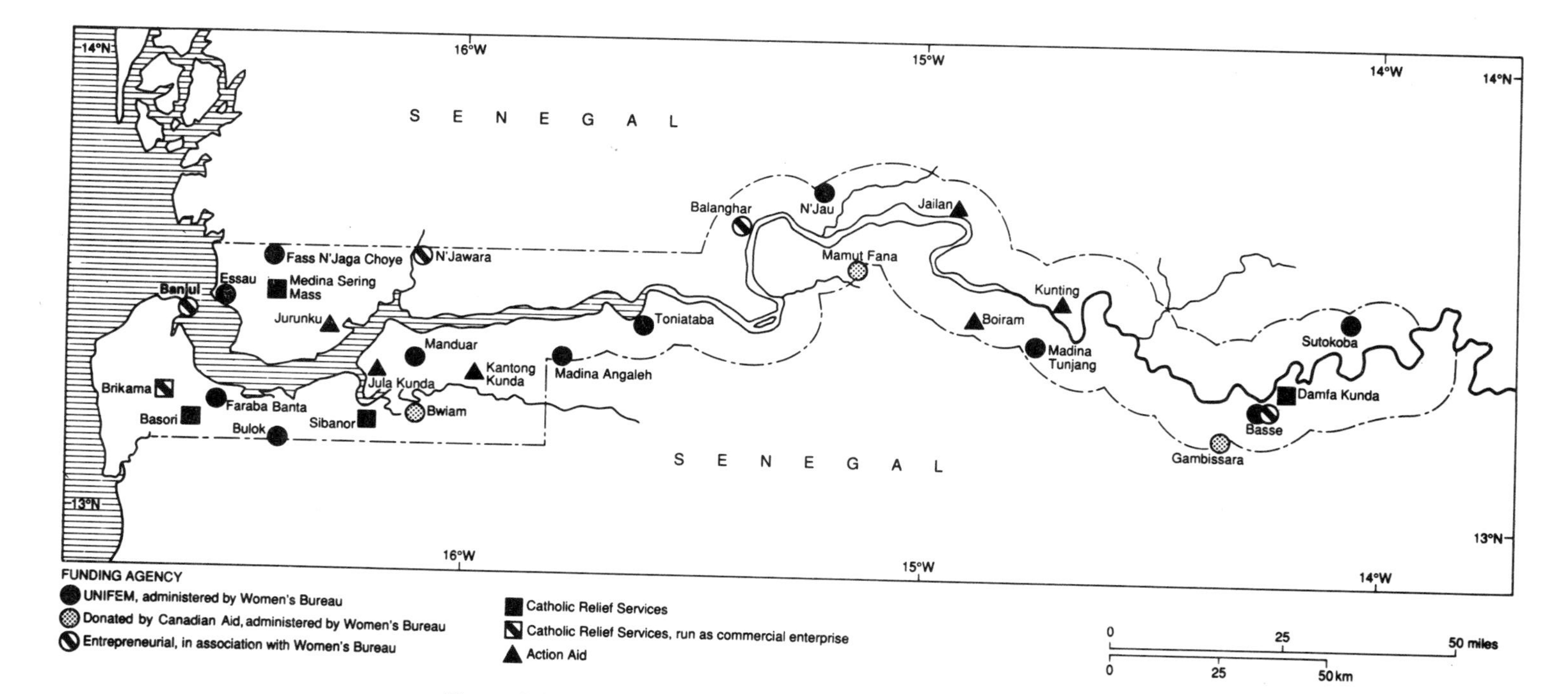

Figure 5.1 Location of *coos* mills in The Gambia, August 1988

time; control of the technology; acceptability; and sustainability of the technology. The village, with a population of 1,416, is on the main road from Banjul to Casamance (Senegal), 50 km from Banjul. The village has few development inputs: a monthly clinic, a Maternal and Child Health (MCH) programme operated by CRS and a sealed, hand-pumped well donated by the Saudi–Sahelian Project. The nearest primary school is 6 km way at Jiboroh, as is the agricultural extension worker. The main crops grown are groundnuts, *coos* and maize (by men) and rice (by women).

Access

The *coos* mill at Basor, donated by the Catholic Relief Service (CRS), was opened in November 1986 and designed to serve the population of Basori and ten small neighbouring villages. The machinery consists of a hammer mill and huller powered by a diesel unit. It is housed in a purpose-built secure shed at one end of the village and is open each evening from about 5 to 7 p.m. The authors' survey showed that within the village, access is universal, with every household making use of the machine. The main constraint on daily usage is the availability of ready cash. Use fluctuates seasonally with the availability of *coos*: more use is made of the mill during the post-harvest months December to July. Milling charges are affordable (10 pence for 4 kg) and most women meet the charge from their own (51 per cent) or a combination of their own and their husbands' income (49 per cent). The use of husbands' money is interesting, illustrating that men also value the utility of the mill. Our research found that women are not expected to work on their husbands' farms to compensate for this cash payment, as was found in a similar study in Senegal (Carr 1985: 128).

The mill is not used by the neighbouring villages, as was intended by CRS, partly because of distance from Basori. Discussions with local women revealed that they are prepared to spend no more than 30 minutes walking to the mill; all neighbouring villages are beyond this range. In addition, there are political rivalries between the villages: neighbouring settlements are apparently unhappy that the mill was sited at Basori. For these reasons the Basori mill has low overall usage rates, affecting its profitability.

Time and energy saved

Preparation of *coos* involves three stages: threshing, winnowing and milling. Mechanised mills can potentially replace the second and third stages of this process with a huller and hammer mill, but not the first. It is important to note, therefore, that the potential saving of time is not

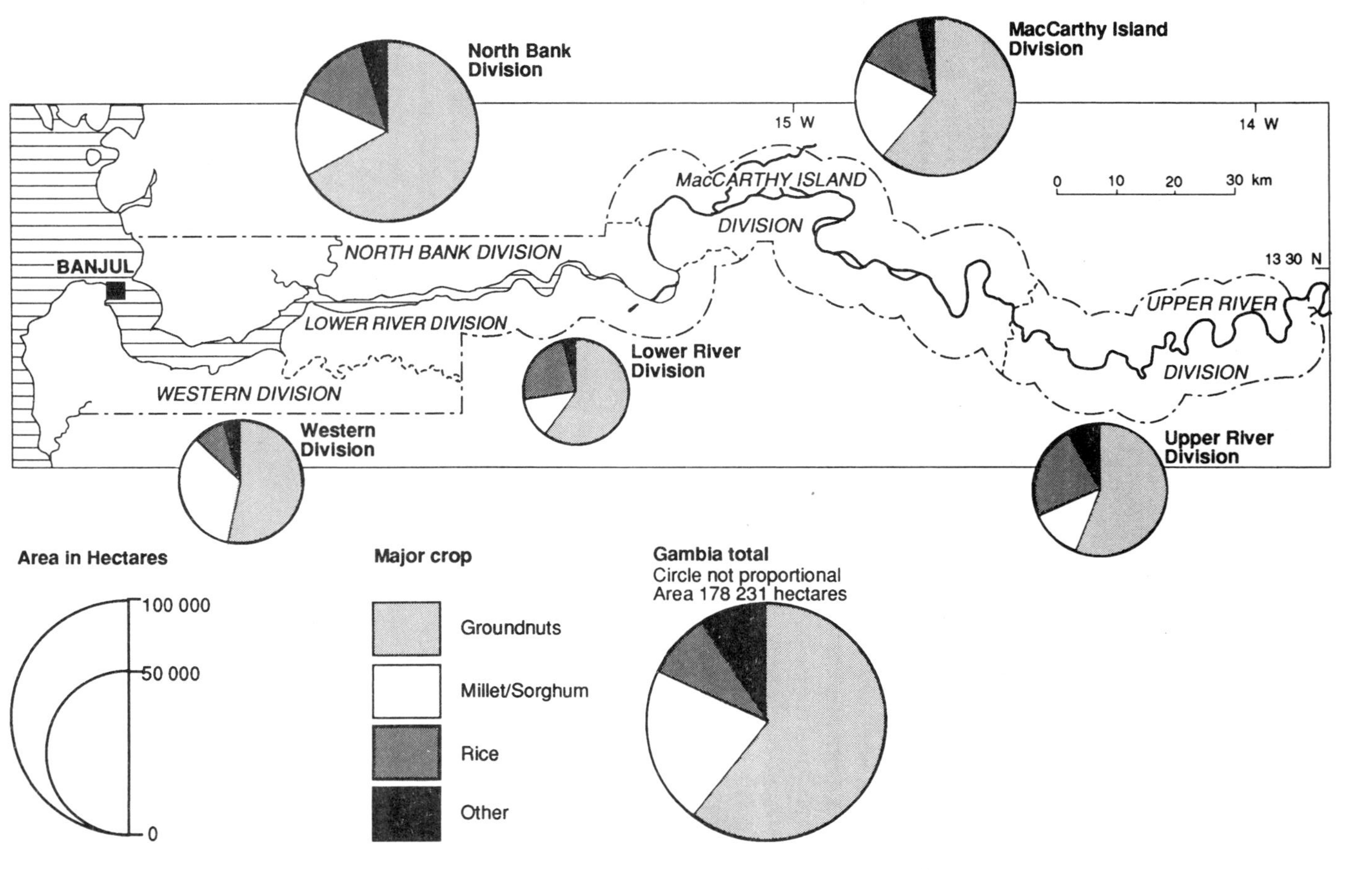

Figure 5.2 Area under cultivation by major crops, 1976–7

the total 4 hours of the production process, but 2–3 hours, depending on the amount of meal or flour required. This is less than many recent reports have suggested (UNIFEM 1988).

Without exception, women in the village pointed to the saving of energy as a more important benefit than saving time. Pounding cereal to flour using a pestle and mortar is hard physical work, and women appreciated being released from this task. Unlike savings in time, saving of energy cannot readily be quantified, but there can be no doubt that it contributes to the women's health and well-being and the more efficient performance of their other productive and reproductive tasks.

Calculation of the exact time saved by women in using the mill is complicated because households are polygynous and differ in their organisation of meal preparation and hence the task of milling. The average time saved by each woman is no more than one to one and a half hours per day, because most households have a rota system within the cooking unit (*sinkiro*) giving women days on and days off cooking duties. Those on duty may save 2–3 hours per day by using the mill (Table 5.2), but not everybody saves a block of time each day. It should

Table 5.2 Typical rainy season time budget for rural Gambian women

Time	*Without* coos *mill*	*With* coos *mill*
0600	Wake, wash, dress Prepare and give children breakfast Sweep compound Prepare lunch (rice) Do laundry	
1000	Pound *coos*	Leave for fields
1100	Leave for fields Take rests during day, for eating, breastfeeding, etc. Collect bush products	
1630	Leave fields for return to village	
1730	Pound and prepare evening meal	Leave fields for return to village
1830		Go to mill Prepare meal Rest/chores
1930	Eat evening meal	Eat evening meal
2100	Bath and retire	Bath and retire

Saving in time: 1 hour in morning
1 hour and 30 minutes in evening: 1 hour extra in fields, 30 minutes in compound.

Source: Fieldwork, 1988

be noted however, that the members of the cooking rota are the younger women of the household who are actively involved in child rearing and household and agricultural production. Any time saved is therefore valuable. In many households, children are sent to the mill, thereby saving women the walking, queueing and waiting time. In this way, family labour is utilised more efficiently.

The use made of this time could not readily be attributed to any particular task by the women. Most said they either spent longer in their rice fields – thus increasing food production – or had more time to spend on their other household jobs – thus contributing to household welfare (Table 5.2). Nobody in the survey identified any new activities now undertaken in the time saved, perceiving it to be very little. Development activities have not been introduced in conjunction with labour-saving devices, although women would now have marginally more time to participate.

Control

Milling machines in rural areas of The Gambia, as in Basori, are managed by a village committee with a woman president and predominantly female members. In Basori, the two part-time mill operators are men, and so too is the treasurer/book-keeper, because there are no literate women in the village. The machines were a gift from the donor agency (CRS) and control ultimately rests with them. Thus women have become involved in the management of the mill but are constrained from taking full control by their lack of literacy. The Gambian national female literacy rate is 15 per cent (UNCTAD 1988) but rural literacy rates are virtually zero. Our research found that only 2 per cent of women in Basori had ever attended school, and none for longer than 3 years.

Acceptability

The *coos* mill is much appreciated by the villagers – men and women alike. The milled flour is of good quality and the energy and time saved by women are highly valued. The hulling machine, however is underutilised. Women complain that it does not remove the hull as efficiently as their traditional method of soaking, an activity which is not demanding of time or energy. Also, the huller costs an additional 10p per 4 kg load, making the total costs unacceptably high and was hampered by frequent break-downs.

Sustainability

Intermediate Technology (1985) estimate that the real annual running costs for the first 5 years of a grain hammer mill costing £2,000 are on average £6,130. Of this, £380 is loan repayment on the capital, £400 depreciation, £350 for repairs, £3,000 for fuel and £2,000 for labour (Intermediate Technology 1985: 145). The mills in The Gambia are all gifts from donor agencies, therefore the major recurring costs are maintenance, fuel and labour. However, the mills are expected to make an operating profit and build up a revolving fund sufficient to cover major repairs and a replacement machine after a 5–year write-off period.

Even taking account of lower operating costs than the IT example, it is quite clear that on economic criteria, the mills at Basori, and those in most other village communities, are not sustainable. The capital costs of the machinery and all the training and support services are supplied as an aid package, but the running costs are supposed to be met from the mill's operating profits. Village records for Basori mill are incomplete but CRS considers that the machine makes an operating loss because throughput is too low. This is partially explained by the low user population, but also by seasonal fluctuations in use. Seasonal variation in the availability of local *coos* and seasonal variations in money supply associated with the sale of cash crops coincide to give greater usage of the mill from December to July. In addition, the food aid given by CRS as part of its MCH nutrition supplementation schemes (4 kg of cereal once a month to 50 participating mothers) is polished rice, not *coos* (CRS 1988a). This is cooked without further processing, thus affecting demand for grain-milling services.

Figure 5.3 depicts the cumulative savings (at August 1988) of the UNIFEM mills administered by the Women's Bureau, from which it is clear that only the mills at Faraba Banta and N'Jau have in excess of D11,000 (£1000), after 3 years of operation. Catholic Relief Services report that of the five mills donated, one urban and one rural centre make a profit, one breaks even and two make an operating loss (CRS 1988b), demonstrating that machines of this size and level of technology are seldom sustainable by village host communities, a fact now recognised by CRS. On welfare criteria, however, the mills make a significant positive impact. Thus, where aid involvement cushions the villagers from the economic losses of the project, a strong argument could be made for supporting milling machines on welfare grounds alone. However, this contributes to the perpetuation of dependency and lack of village autonomy. A more appropriate machine, with smaller throughput, hand powered and less expensive to run, would enable villages to be self-supporting and women to have greater control of the technology.

CONCLUSION

This research shows that these labour-saving devices do indeed alleviate women's workloads, but more by the saving of energy than the gaining

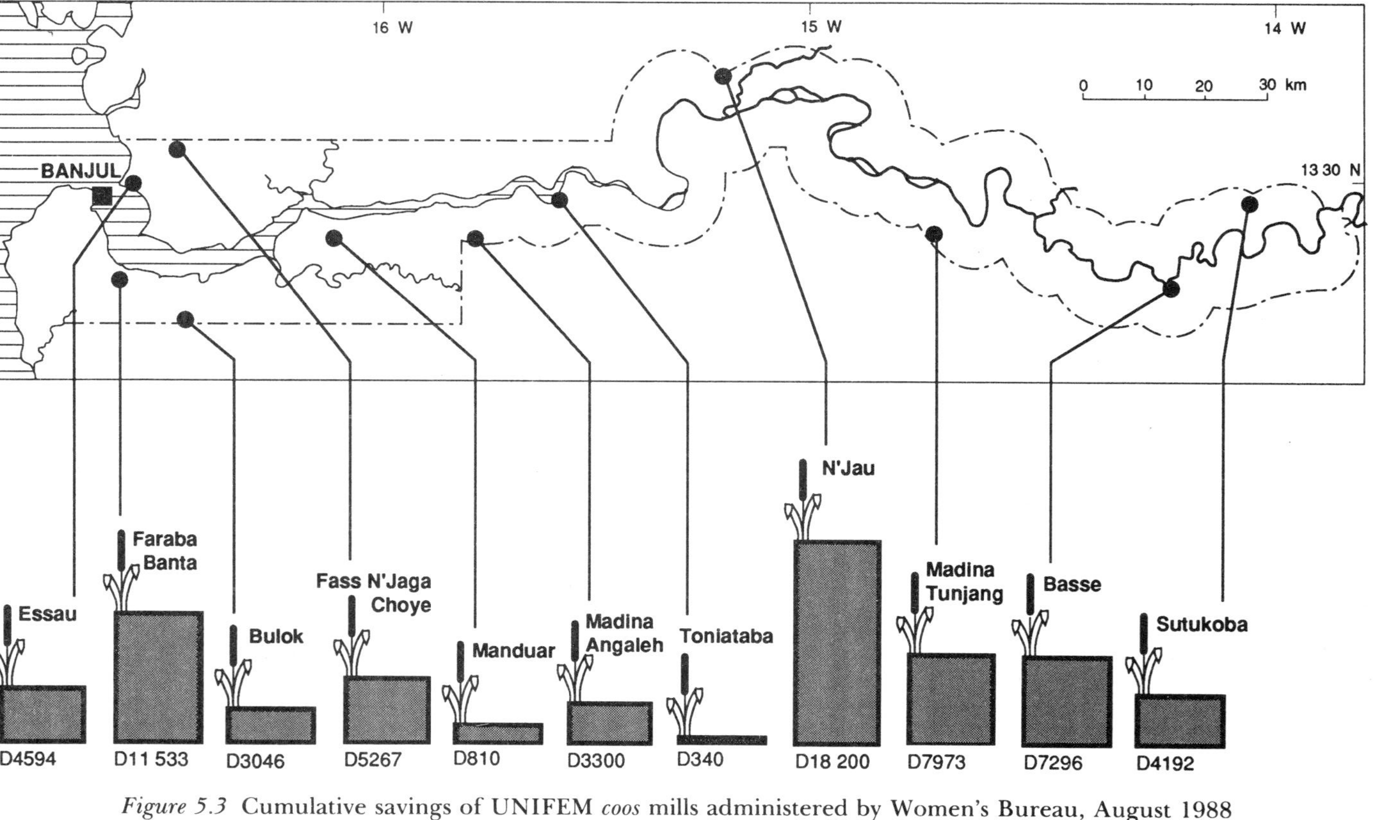

Figure 5.3 Cumulative savings of UNIFEM *coos* mills administered by Women's Bureau, August 1988

of time. The machines do not, by themselves, permit women to participate more fully in the development process but they are an essential first step to raising labour productivity. They also save the energy and time of the most hard-pressed group of Gambian women: those in their reproductive years whose contribution to household and agricultural production is central to the development of their communities. However, women's full participation in the development process is still constrained in The Gambia, as in many countries in sub-Saharan Africa, by insufficient attention to women's agriculture (Trenchard 1987) and very low levels of female literacy. Until these vital issues are acknowledged and addressed as central parts of government development strategy, the full potential of improved technologies for rural development cannot be realised.

6

THE IMPACT OF CONTRACEPTIVE USE AMONG URBAN TRADERS IN NIGERIA

Ibadan Traders and Modernisation

B. Folasade Iyun and E. A. Oke[1]

INTRODUCTION

Concern over levels of contraception in African countries has assumed a new dimension in recent years due partly to phenomenal population growth rates and the depressing economic situation in most of these nations. This interest can also be attributed to the work of international agencies with health and family planning (FP) services at the community level, coupled with recent positive changes in government policy towards population issues in many African countries, including Nigeria.

The percentage of women aged 15–49 utilising contraception in Africa has been found to increase significantly as information and education reach more eligible women, yet wide variations from country to country and between regions remain. Within Africa, the percentage of women aged 15–49 that utilised FP ranged from 5 per cent (Lesotho) and 6 per cent (Nigeria) to 29 per cent in Zimbabwe and 53 per cent in Mauritius (IPPF 1985). Such variations in the extent of contraceptive usage are closely associated with 1) awareness, 2) realisation of the cost of rearing children and 3) the quality of life desired for one's children (Farouq and Adeokun 1976). Family-planning programmes often have rapid impact. For instance, in Ishan Bendel State, Nigeria, the percentage of women with FP knowledge in 1969 was less than 1 per cent before a rural FP programme was introduced. This figure jumped to 25 per cent in 1972 (Farouq and Adeokun 1976). Similarly, in rural areas of Oyo State, a rise in contraceptive acceptance from 2 per cent to 15 per cent was observed within 12 months of the inception of a rural FP programme.

The place of residence of women no doubt exercises a profound influence on the knowledge and practice of contraception. In Jordan, it was observed that only 1 per cent of urban dwellers were ignorant of family planning methods, but in the semi-urban and rural areas, the figures were 2 per cent and 11 per cent respectively (Rizk 1977). Meanwhile, significant differentials were found in current use with the rural areas recording 6 per cent, the semi-urban 24 per cent and the urban centres 38 per cent. In a recent study it was observed that within the urban area of Ilorin, Nigeria, the place of residence was statistically significant in the pattern of contraceptive use, in addition to education and age-factor variables within a stratified general population (Oni 1986).

The central concern of this chapter is two-fold: 1) to highlight the trend of contraceptive use in the second-largest city in Black Africa and 2) to ascertain the existence of spatial variations in contraceptive use among a particular low-income group: female market traders. It is expected that the information generated will lead to an improvement in contraceptive take-up in the urban setting.

The choice of market traders as the subjects studied in this paper has been influenced by the following factors. Firstly, markets play a very significant dual role in the social and commercial life of Africans and, in particular, Nigerians. They serve as hubs of economic, cultural, recreational, political and ritual activities in the African landscape (Vagale 1972). Secondly, trading is very popular among African women, especially those with little or no education (Farouq and Adeokun 1976). In particular, market women form a formidable female workforce in the urban setting even though many of them belong to the low-income group. Also, it has been discovered that market traders have high patronage (49 per cent) of contraception among Lagos clients (Oyediran and Ewumi 1976).

THE BACKGROUND AND GROWTH OF FAMILY PLANNING SERVICES

Due to an increase in the number of abandoned babies in the late 1950s, the Family Planning Council of Nigeria (now the Planned Parenthood Federation of Nigeria) was established. It set up the first FP clinic in the country in Lagos in 1958 which was followed by one in Ibadan in 1964 (Ojo 1984). In 1965, the FP clinic at University College Hospital, Ibadan, became an independent clinic under the direction of Professor P.A. Ojo. The clinic was sponsored by the Population Council of New York. It also had branches at Adeoys State Hospital and Inalende Maternity Centre owned by the Municipal Government. Today, four other clinics exist providing services similar to those offered in private clinics.

By 1970, 1,940 clients were registered in the clinic, but within three years, this figure had increased by over 80 per cent (Table 6.1). In 1976, there was again a substantial increase, but the figures dropped during the next two years and picked up again in 1979. Table 6.1 depicts the trend in growth of family planning clientele from 1970 to 1988. The increase which began in 1982 has been maintained to date. This cannot easily be explained by a substantial expansion of knowledge of family planning services at the University College Hospital as more clinics have

Table 6.1 Growth trends of family planning clients, University College Hospital, 1970–88

		Type of contraception					
Year	*Total new clients*	*Pill*	*% of total*	*IUD*	*% of total*	*Other*	*% of total*
1970	1,940	100	5.2	1,260	64.9	580	29.9
1971	2,250	440	19.6	1,720	66.4	90	4.0
1972	2,140	610	28.5	1,450	67.8	80	3.7
1973	2,264	660	29.2	1,600	70.7	4	0.1
1974	1,640	360	22.0	1,280	78.0	0	0.0
1975	2,140	780	36.4	940	43.9	420	19.6
1976	3,050	1,060	34.8	1,260	41.3	730	23.9
1977	2,100	790	37.6	1,180	56.2	130	6.2
1978	2,180	690	31.7	1,380	63.3	110	5.0
1979	2,590	700	27.0	1,700	65.6	190	7.3
1980	2,640	495	18.8	1,620	61.4	525	19.9
1981	2,220	320	14.4	1,860	83.8	40	1.8
1982	2,950	420	14.8	2,380	80.7	150	5.1
1983	3,700	420	11.4	2,540	68.6	740	20.0
1984	3,611	370	10.2	2,701	74.8	540	15.0
1985	4,192	432	10.3	3,370	80.4	390	9.3
1986	4,835	422	8.7	3,586	74.2	827	17.1
1987	4,963	380	7.7	3,685	74.2	888	17.9
1988	4,274	403	9.4	3,442	80.5	429	10.0

Source: Fertility Research Unit, University College Hospital, Ibadan 1989

Table 6.2 Trends in attendance at obstetrics and maternity clinics at the University College Hospital, 1978–88

Year	*Total obstetrics*	*Total maternity*
1978	4,328	22,398
1979	3,913	20,127
1980	3,066	18,685
1981	3,781	23,204
1982	3,780	25,406
1983	3,124	18,167
1984	3,531	19,999
1985	2,927	47,746[a]
1986	2,170	13,415
1987	1,969	11,482
1988	1,754	11,221

Source: The University College Hospital, Ibadan, 1989
Note: [a] Dramatic increase

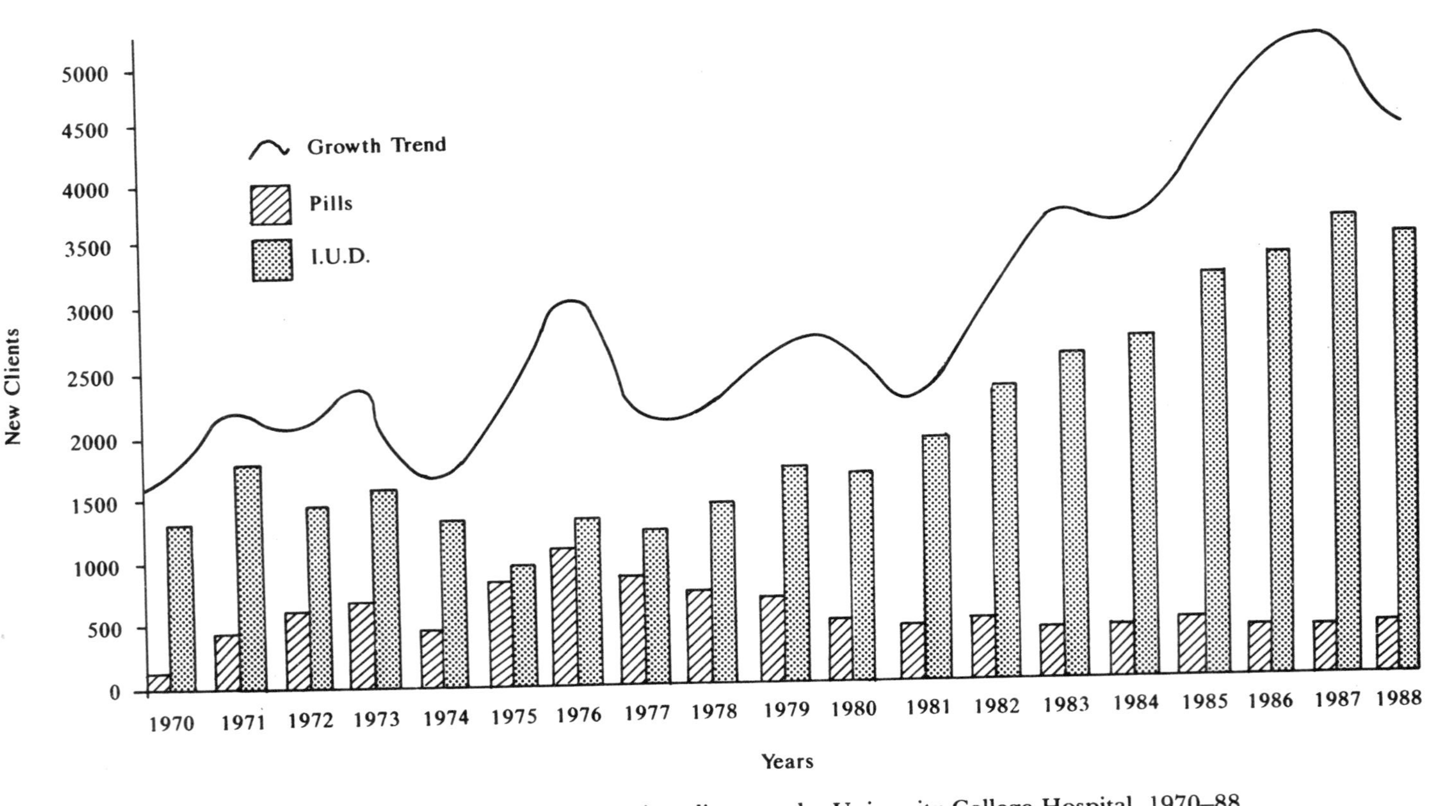

Figure 6.1 Growth trends of family planning clients at the University College Hospital, 1970–88.

now been established within the city. While family planning clientele increased substantially at University College Hospital, maternal attendance at the Department of Obstetrics and Gynaecology decreased (Tables 6.1 and 6.2).

At this juncture, some inferences on the general trend of family planning clientele can justifiably be made. The sudden fall in patronage in 1974 could possibly be due to the immediate impact of the oil boom when the oil price suddenly jumped from US$2.00 to US$14.00 per barrel. In 1975, there was a change of government which ushered in mass retrenchment and early retirements and created a state of uncertainty until late 1976.

However, in late 1976, Universal Primary Education was introduced by the federal government which could have further reduced the clientele in 1977 and 1978. In 1979, the impact of the then new military government austerity measures could have forced more women to take to family planning. By 1981, many Nigerians erroneously thought that the economic situation had stabilised and this probably explains the decline in the patronage of family planning services at the clinic. But, by 1982, the harsh economic realities had dawned on most Nigerians, hence the trend from 1982 till the present. Relatively high figures were recorded compared with the 1970s, for other maternal clinic attendance.

The euphoria of having a new regime in 1984 probably made family planning less attractive than the previous year. But as the political and economic atmosphere darkened more women embraced the scheme in 1985. The Structural Adjustment Programme (SAP) introduced in 1986 cast an air of gloom if not doom and even more women found solace in contraception as insulation against the general hardship which has brought inflation of the order of 500–1,000 per cent. Significantly, even though erroneously, as the positive effects of structural adjustment were becoming apparent in 1988 the figure dropped again. Indeed, the specific cause here may have been the review of salaries and fringe benefits which reflected the economic situation temporarily. A four-period moving average is used to estimate the trend, given the behaviour of the new client curve of Figure 6.1 which seems to have a new crest every four years.

Table 6.1 depicts an increasing trend beginning about 1978–9, the year Nigeria first went into recession as a result of the glut in the world oil market. The boom years 1974–77 show a slightly falling trend which is close to a constant average trend. The years in which an increasing trend is much more observable coincide with the 1980s when Nigeria was rocked by external shocks resulting from global recession and the perpetual glut in the world oil market. Thus, there seems to be a strong linear relationship between the number of new family planning clients and economic deprivation.

Table 6.3 Number of patients treated for nutritional diseases, by sex

	1983		*1986*	
Month	*Male*	*Female*	*Male*	*Female*
January	–	–	60	49
February	–	–	68	47
March	98	106	35	29
April	60	50	83	51
May	76	68	65	57
June	136	131	106	63
July	104	91	70	50
August	113	104	67	37
September	115	130	107	58
October	124	110	67	60
November	106	119	79	41
December	98	125	84	58
Total	1,030	1,034	891	600
% of total treated	49.9	50.1	60.2	39.8

Source: University College Hospital, Ibadan, 1989

There is no doubt that the present level of socioeconomic development in the country is playing a more significant role in the attitudes of Nigerian women towards the need to utilise family planning at least to space births if not to limit the number of children born. Simply put, the economic depression that has followed the debt crisis since 1983 in particular seems to encourage responsible motherhood. As structural adjustment policies continue to diminish the quality of life of the most helpless segments of the Nigerian population, mothers and children, more and more women find solace in family planning services.

On the other hand, a relatively sharp decline in clinic attendance has been noticed in many medical institutions in the country (Saba 1987). Indeed, the attendance figures at the University College Hospital amply illustrate this general trend. Table 6.3 illustrates a notable decline in the clinic attendance of females treated for nutritional diseases in 1983 and 1988. As depicted in Table 6.3, both males and females contributed almost equally to the clinic attendance in 1983. Table 6.2 further highlights this observation. While the attendance at both the obstetrics and maternity clinics decreased from 1983 (except in 1985), the family planning clientele in Table 6.1 has increased substantially during the same period.

The decrease in the former can be explained by the sharp increase in user charges for confinement and delivery (Osuntokun 1987) which have risen eight-fold in the past few years. Thus, the women in Ibadan are becoming more amenable to modernisation in respect of adoption of family-planning services. However, this cannot be said to be true throughout the city. As depicted in Figure 6.1, more and more women have adopted the IUD since 1982 compared with the number using the Pill. It seems the former fits more readily into the over-burdened programme of activities of many Nigerian women compared with the regimented compliance of the Pill.

KNOWLEDGE, ATTITUDES AND PRACTICE (KAP) OF FAMILY PLANNING AMONG MARKET WOMEN OF IBADAN

The KAP survey was conducted to provide background information on existing levels of contraception use among market traders, in spite of the seemingly gradual increase of family planning clientele in the clinics of Ibadan city. Since the traders to a large extent, represent the low-income groups (urban poor) and were the target of the health programme, it was necessary to ascertain the state of affairs among them.

The Nigerian Fertility Survey (1984) revealed that contraception use was very low among Nigerian women. Also, the level of awareness was said to be low. The statistics indicated that two out of every three women claimed ignorance of family planning, 84.9 per cent never used family planning while only 14 per cent reported they had used either modern or traditional methods; only 6 per cent were users at the time of the survey. However, levels of use in the cities were said to be much higher

Table 6.4 Approval of family planning methods (by market)

	Total		*Approve*		*Maybe*		*Disapprove*	
Market	*£*	*%*	*£*	*%*	*£*	*%*	*£*	*%*
New markets								
Oke Ado	90	100.0	53	59.0	23	26.0	14	15.0
Dugbe	204	100.0	148	73.0	20	10.0	36	17.0
Eleyele	112	100.0	78	70.0	14	13.0	20	17.0
Apata	39	100.0	28	72.0	7	18.0	4	10.0
Sango	117	100.0	78	67.0	9	8.0	30	25.0
Mokola	55	100.0	32	58.0	5	9.0	18	33.0
Total	617	100.0	417	67.6	78	12.6	122	19.8
Old markets								
Agugu	43	100.0	30	69.0	9	22.0	4	9.0
Owode	59	100.0	40	68.0	10	17.0	9	15.0
Oranyan	62	100.0	42	68.0	17	27.0	3	11.0
Elekuro	25	100.0	19	76.0	6	24.0	0	0.0
Ibuko	44	100.0	34	77.0	10	23.0	0	0.0
Ode Oolo	34	100.0	5	15.0	10	29.0	19	56.0
Total	267	100.0	170	64.0	62	23.0	35	13.0

Source: Fieldwork, 1988

than that in the rural areas. Dow (1977) reported that 62 per cent of women in Ibadan indicated that they had taken measures to delay early pregnancy. Nonetheless, abstinence appears to have been preferred for birth spacing by the majority of the urban poor. It is appropriate therefore to actually estimate levels of acceptance and use of family planning methods among specific target populations in Third World cities, instead of merely assuming a general level of high acceptance in such urban centres.

To a large extent, the geographical locations of the markets reflect the residential areas of the traders (Figure 6.2). Traders in the modern section are often more heterogeneous and have wider residential coverage in the city. On the whole, 66 per cent of the traders interviewed approved of family planning, 16 per cent were undecided and 18 per cent disapproved. However, approval does not necessarily imply utilisation. It is interesting to note that, between 58 and 78 per cent of the traders in the markets in the modern section approved of family planning (Table 6.4).

Table 6.5 reveals a high level of awareness of family planning among the traders, except at Ode Oolo. The idea of family planning seems to be generally accepted by the traders even though the clinics are underutilised. Table 6.6 shows that only about 33 per cent of the traders admitted ever using modern family planning methods at one time or another. Again, spatial differentials were noticed among the markets.

Table 6.5 Knowledge of family planning methods (by market)

	Total		*Yes*		*No*	
Market	*No.*	*%*	*No.*	*%*	*No.*	*%*
New markets						
Oke Ado	90	100.0	69	77.0	21	23.0
Dugbe	212	100.0	135	64.0	77	36.0
Eleyele	115	100.0	91	79.0	24	21.0
Apata	35	100.0	26	74.0	9	26.0
Sango	117	100.0	97	83.0	20	17.0
Mokola	55	100.0	33	60.0	22	40.0
Total	624	100.0	451	72.3	173	27.7
Old markets						
Agugu	43	100.0	39	91.0	4	9.0
Owode	56	100.0	56	100.0	0	0.0
Oranyan	62	100.0	49	79.0	13	21.0
Elekuro	25	100.0	22	88.0	3	12.0
Ibuko	44	100.0	36	82.0	8	18.0
Ode Oolo	36	100.0	12	33.0	24	67.0
Total	266	100.0	214	80.5	52	19.5

Source: Fieldwork

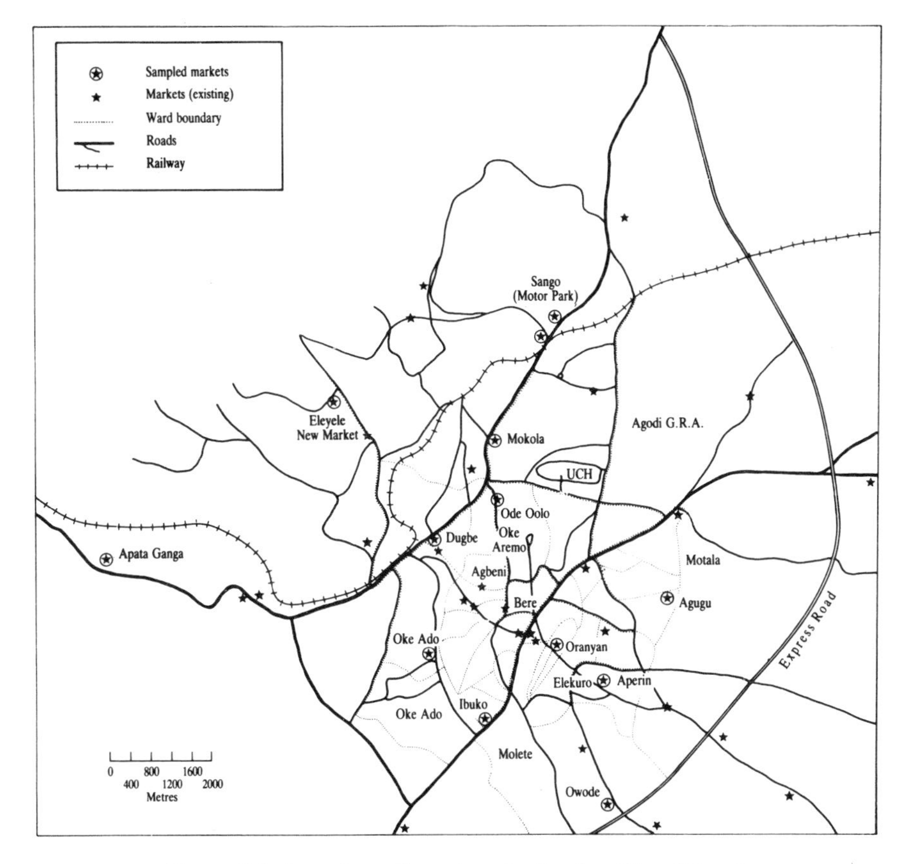

Figure 6.2 Location of markets in Ibadan City, Nigeria

Most of the traders whose markets are located in the core area and who reside mostly in the same neighbourhood, namely Elekuro, Ode Oolo and Oranyan (6 per cent), have never used family planning methods.

Even though 33 per cent of the traders confirmed their use of family planning, only 24 per cent of them indicated current use of the clinic at the time of the survey. This rate is no doubt low by Western standards. Nevertheless, it represents a great improvement on the national usage level. It also appears that abstention as a family planning technique is becoming less widespread among the traders in Ibadan City. Only 26 per cent reported the use of abstention for birth spacing. As indicated in Table 6.6, the traders in the traditional section of the city such as Oranyan, Ode Oolo, Ibuko and Elekuro reported they were not using any modern family planning method at the time of the survey.

Table 6.6 Current usage of family planning methods (by market)

Market	*Total*		*Yes*		*No*	
	No.	*%*	*No.*	*%*	*No.*	*%*
New markets						
Oke Ado	57	100.0	12	21.0	45	79.0
Dugbe	166	100.0	40	24.0	126	76.0
Eleyele	78	100.0	22	28.0	56	72.0
Apata	67	100.0	11	16.0	56	84.0
Sango	64	100.0	21	33.0	43	67.0
Mokola	26	100.0	3	12.0	23	88.0
Total	458	100.0	109	23.8	349	76.2
Old markets						
Agugu	32	100.0	11	34.0	21	66.0
Owode	53	100.0	15	28.0	38	72.0
Oranyan	12	100.0	0	0.0	12	100.0
Elekuro	4	100.0	0	0.0	4	100.0
Ibuko	8	100.0	0	0.0	8	100.0
Ode Oolo	24	100.0	0	0.0	24	100.0
Total	133	100.0	26	19.5	107	80.5

Source: Fieldwork, 1988

It is obvious that the geographical location of the markets seems to influence the level of contraception use. Since markets serve as centres of economic, social and cultural activities, they also help to disseminate information on innovation, such as family planning. They equally possess hierarchical arrangements in terms of interaction and other activities. One could expect the traders in higher-order markets such as Dugbe, which are more heterogeneous in population, to be more amenable to social changes as in the case of adoption of family planning techniques. As expected, markets located close to the commercial business district (Dugbe) have wider catchment areas for the traders. On the other hand, markets sited in or near the traditional core area are more homogeneous in population and are conservative about adopting family-planning practices.

Besides the geographical location of the markets, other factors such as religion and education also influence the levels of contraception among the market traders. The analysis revealed that 44 per cent of Christians indicated that they have used the clinic as compared with only 23 per cent of Muslims. Finally, there is a positive relationship between education and family-planning usage. A sharp difference is observed in the level of contraception use between illiterates and those who had primary education (Table 6.7). The traders found in the modern section of the city were often better educated and by and large more enlightened compared with those found in the traditional areas of the city.

Table 6.7 Use of family planning methods (by level of education)

Use family planning	Education level			
	Total	*None*	*Primary*	*Secondary*
Yes	121 (34.1%)	39 (22.9%)	65 (44.8%)	17 (43.6%)
No	233 (65.8%)	131 (77.1%)	80 (55.2%)	22 (56.4%)
Total	354 (100%)[a]	170 (100.0%)	145 (100.0%)	39 (100.0%)

Source: Fieldwork, 1988
Note: [a] Some totals are rounded

CONCLUSION

It has been noted that increased levels of awareness encourage more eligible women to adopt contraception. The gradual, and in some cases, sharp increase in the family-planning clientele gives credence to this observation. Changes in economic conditions also often lead to subsequent demographic and social adjustments, often referred to as 'survival strategies'.

In Ibadan, the analyses indicate clear increases in the use of contraception, especially in recent years, partly as a response to the economic crisis being experienced by the country. This is evident from clinic data at the University College Hospital. The increase in family planning contacts made by the Community Based Delivery (CBD) agents, who were trained to operate in the city markets, also confirms this observation.

There is abundant evidence to show that even low-income groups, such as market traders, are fast adopting family-planning methods at least to space births. Nonetheless, differentials are noticed in the level of usage from market to market, which is partly a reflection of their geographical location. More noticeable increases in levels of contraception were observed in markets located in the modern section of the city which are often more heterogeneous in population characteristics. Thus, for planning purposes, disaggregation of the target population may be necessary instead of just assuming homogeneity when introducing an innovation in a Nigerian city or perhaps any city in a developing nation.

Part III
SOUTH ASIA

INTRODUCTION

Women and gender in geography

An overview from South Asia

Saraswati Raju

Gender is emerging as a significant research area in the essentially androcentric discipline of geography. However, taking stock of the place South Asian geographers occupy in this exciting field is a somewhat piquant exercise. Although scholars, particularly Indian ones, take the lead in feminist research, the share by fellow geographers is relatively minuscule. Why this is so is an intriguing question. This note is an attempt to address this question while presenting an overview of the existing research and trends.

The search for a geographical paradigm from logical positivism to humanistic overture is too well-known to necessitate elaboration. What is important to note is that even the latter shift failed to produce much insight into the lives of women and the concern therein for the 'full range of human experience' proved to be really the concern for men who usurped the focus. The treatment of women at best remained incidental, peripheral or subservient to that of man. This intellectual aberration was 'imported' into South Asia from the Western world rather uncritically. The point of contention, however, is that while in other countries geographers have started taking a pointed interest in issues relating to gender, the South Asian geographers, in general, lack a coherent effort in this direction, let alone an official forum for recognition.

It was not until very recently that gender became an issue for serious analysis. In South Asia, as elsewhere, the concern for women as a subject for study arose as a part of larger issues related to socio-spatial disparities and distributive justice within the rubric of social geography, although isolated studies can be traced back to the 1960s (Gosal 1961; Chadna 1967; Mukerji 1971). However, this interest in gender is pursued only by a few individual scholars rather than as a part of collective efforts to produce a feminist perspective in geography.

This dubious status is essentially because of the still-existing traditionally constructed paradigms and ideas about what geography addresses and/or ought to address, alternatively, deciding what could or could not be accommodated by the subject. No wonder that when the Indian Council for Social Science Research (ICSSR) decided to commission a series of studies related to the status of women to help the National Committee on the Status of Women (set up in 1971), its task force had anthropologists, political scientists and sociologists, but no geographers! In 1982–3, separate fellowships were instituted by the ICSSR for women's studies. However, all such scholarships have been taken by disciplines other than geography. Reports from other South Asian countries are not drastically different.

A highly structured curriculum in most of the universities in the region does not permit the incorporation of women as an independent subject for geographical enquiry, although some isolated courses may include women as incidental to the development of some other perspective. There is, however, one exception to this as the multi-disciplinary Centre for the Study of Regional Development (CSRD) at Jawaharlal Nehru University, New Delhi, recently introduced an M.Phil-level course on the 'regional dimension of the female labour force', the first course focusing on gender to be taught in geography in India.

The centre has also undertaken several studies on gender issues by geographers, demographers and regional economists. These can broadly be grouped into two categories: (1) those where women's concerns are addressed in so far as they contribute to an understanding of regional inequalities (see Mitra 1978, 1979a, 1979b; Chopra 1979; Kundu 1986) and (2) those which deal directly with geographical analyses of gender issues (see Raju 1981, 1982, 1984, 1987, 1988). To date there are nine M.Phil dissertations and three doctoral theses (two completed and one in progress) which deal explicitly with gender. Outside CSRD, three dissertations on women have been submitted in the department of Geography at Bombay University, whereas in Osmania University at Hyderabad, an action research project on women's cooperatives has been undertaken by geographers.

Since in Indian geography explicit emphasis on gender is very recent, the volume of research is small. A detailed review of this literature is not attempted here due to space constraints. However, in general, the thrust is on identifying patterns of literacy, work participation, fertility, sex ratios and migration (Nagia 1983; Nagia and Samuel 1983; Nayak and Ahmad 1984; Swarnkar 1985; Mittal 1986; Satish 1984; Nuna 1986; Patnaik 1989).

As stated earlier, the concern for women actually arose as a concern for distributive justice whereby women emerged as a deprived segment of any given population. A logical approach to the whole question in such situations has, therefore, been to see whether the deprivation of women forms a part of an overall deprived scenario, or their deprivation is qua women. This brings in a complex interplay of class, gender and caste (in South Asia) in creating observed patterns. With a few exceptions, where explanatory frameworks in terms of various sociocultural and economic variables and their impact on whatever is being studied are provided in order to capture these complexities (Nayak and Ahmad 1984; Raju 1981, 1982, 1988, 1989, 1991), much of the research is of a descriptive nature and is not placed in a proper perspective. Moreover, there exists a heavy reliance on published sources of data. Issues which need field data have largely remained untouched for obvious reasons. A recently concluded project on female labour in rural India by Ahmad (1985–91) is an exception to the rule.

In contrast, several scholars from Bangladesh and Sri Lanka score better in terms of generation of data from field observations although much of such research is likewise confined to issues related to work participation, literacy/education and population dynamics (see Ahsan and Hussain 1987; Elahi 1982, 1987: chapter 7; Ranasinghe 1989; Wickramasinghe 1987a, 1987b). However, more recently, the focus of interest has shifted to issues related to access of females to basic needs, planning and work vis-a-vis development processes (Wickramasinghe, 1990a,b). A recently held seminar on 'women and environment' by the Bangladesh Geographical Society in Dhaka may be indicative of research directions in future.

In sum, although geographical research on women has yet to develop a cohesive feminist perspective and on abiding interest among South Asian geographers, a beginning has been made. That the efforts are gradually gaining ground is evident from a very recently approved Commission on Gender by the National Association of Geographers in India. Besides this, the *Journal of Abstracts and Reviews in Geography* published by the ICSSR has introduced a separate section on research on gender. The future thus seems to be bright, although challenging and a little slower in pace compared to other regions of the world.

7

GENDER RELATIONS IN RURAL BANGLADESH

Aspects of differential norms about fertility, mortality and health practices

K. Maudood Elahi

INTRODUCTION

This paper is an outcome of detailed survey research in two rural areas of Bangladesh on 'Rural Fertility and Female Economic Activity' undertaken during 1983–4 (Elahi 1987). In this study, a 10 per cent sample was randomly selected for a questionnaire survey. This generated a total of 618 households in two rural areas: Baganbari (300) and Kalihati (318) in the district of Chandpur and Tangail respectively (Figure 7.1). The respondents were the wives and their husbands. The survey was also supported by an anthropological in-depth investigation of selected households thereby providing much of the qualitative data for this study (Elahi 1982).

One of the focuses of the research has been to understand differential norms at individual and community levels regarding fertility, mortality and health, and how they relate to family formation and the activity patterns of women at the household level. In this chapter, the social and related norms about fertility, mortality and health practices have been examined and in doing so, some related aspects have also been reviewed.

GENERAL DEMOGRAPHIC PROFILE

The population of both the study areas is predominantly Muslim: 82 per cent and 74 per cent in Baganbari and Kalihati, respectively. The Hindus are the significant minority group in both areas but there is no apparent seclusion of settlements by religion in either of the areas under study.

Fertility levels are generally high in Baganbari and Kalihati, although the latter has a fertility rate that is lower than the former and the

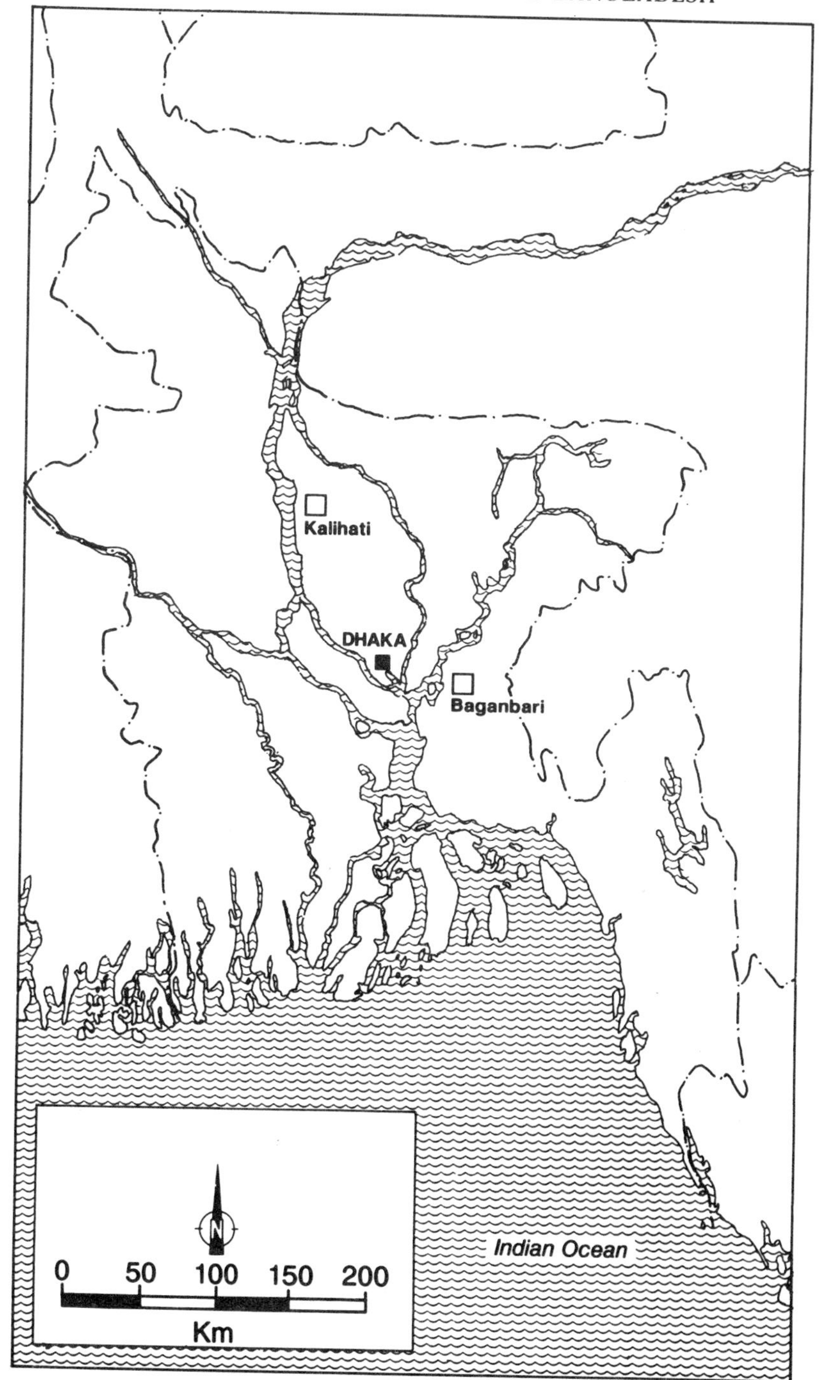

Figure 7.1 Bangladesh: the study villages

Table 7.1 Measures of fertility and mortality in Baganbari and Kalihati, 1983–4

Measure	*Baganbari*	*Kalihati*	*Bangladesh (rural)*[a]
Crude birth rate	40.0	33.0	36.4
Total fertility rate	7.0	6.2	6.2
Crude death rate	15.0	11.0	13.2
Infant mortality rate	114.0	93.0	120.8

Source: Rabbani and Hussain 1984
Note: [a] Refers to 1981–3

Bangladesh national average (Table 7.1). The high level of fertility is not only a major cause of women's high death rates generated from maternal mortality but is also an important variable limiting their lives' options with respect to work and status in society.

The level of mortality is also high in both the study areas (Table 7.1). The infant mortality rate (IMR) is lower in Kalihati than in Baganbari, but both areas are below the national average. The lower IMR in Baganbari may be due to the impact of the health and surveillance facilities available at the International Centre of Diarrhoeal Research/ Bangladesh (ICDDRB) field camp based in the area. The causes of death are varied but diarrhoea and gastrointestinal diseases are the most common in both areas and throughout Bangladesh. Table 7.2 provides the five main causes of death, obtained from the survey results. The responses tally closely with the information collected on the prevalence of diseases, which is discussed later. A change in this pattern can only take place with the improvement of interrelated socioeconomic and medico-environmental conditions. On the other hand, as long as the birth rate remains high, there is a limit to mortality reduction. This view particularly applies to the deaths related to child bearing and infant mortality.

Most inhabitants have a rural origin and approximately 95 per cent of the inhabitants of both villages were born in their respective villages.

Table 7.2 Five major reported causes of death in Baganbari and Kalihati, 1983–4

	% of women respondents	
Causes of death	*Baganbari*	*Kalihati*
Diarrhoeal and gastroenteric	40.0	35.0
Cholera	18.0	20.0
Typhoid	12.0	20.0
Child birth	10.0	15.0
Blood deficiency[a]	20.0	10.0

Source: Author's fieldwork
Note: [a] May be related to other disease such as fever, respiratory ailments and chronic malnutrition

HEALTH INFRASTRUCTURE

Health facilities available in the two study areas do not differ greatly from those found elsewhere in Bangladesh. Kalihati is relatively better off in terms of health facilities, since it is near the *upazila*[1] headquarters, thereby enjoying better access to a number of institutional and administrative linkages. Table 7.3 summarises the health infrastructure in both areas.

Table 7.3 Health facilities available in Baganbari and Kalihati, 1983–4

			Distance to nearest facility					
	Number		*Miles*		*Travel Time*[a]		*Cost*[b]	
Health facility	*B*	*K*	*B*	*K*	*B*	*K*	*B*	*K*
Govt hospital	0	0	10[c]	6[d]	6	6	20	10
Upazila health centre	0	0	10[c]	6	6	6	20	10
Dispensary	1	3						
Maternity centre	0	0	10	6[a]	6	6	20	10
Chemist shop	4	2						
Private doctor	6	4						
Homeopath doctor	3	6						
Paramedic	2	5						
Trained midwife	0	7	5		1		10	
Untrained midwife	6	6						
Family welfare assistant	2	2						

Source: Author's fieldwork
Notes: [a] Travel time is in hours
[b] Cost is in Taka
[c] ICCDRB Hospital in Matlab
[d] Tangail District Hospital

B Baganbari
K Kalihati

During 1983–4, 101 and 128 persons had fallen ill in Baganbari and Kalihati, respectively. The proportional distribution of these persons by duration of illness is shown in Table 7.4. Overall, fever, gastroenteric and diarrhoeal diseases are the most rampant in both the study areas. The treatment pattern adopted by the households is presented in Table 7.5. It can be observed that a relatively small proportion of the sick was reported to have been treated in government hospitals or the *upazila* health centres in both Baganbari and Kalihati. Despite the location of the *upazila* health centre within the study area in Kalihati, this centre was chosen for treatment by a relatively small proportion of the respondents. Private doctors continue to be the preferred source of treatment of sick persons in Kalihati. In Baganbari, most respondents consulted the paramedics for treatment. A number of other sources of treatment was

Table 7.4 Disease pattern by duration of illness of affected persons in Baganbari and Kalihati, 1983–4 (%)

Disease	*Baganbari* *a*	*b*	*c*	*Total*	*Kalihati* *a*	*b*	*c*	*Total*
Fever	67	64	20	47	64	60	41	52
Diarrhoea	20	7	1	9	16	–	2	8
Dysentery	8	14	19	14	5	4	2	3
Pneumonia	2	2	1	2	–	2	1	1
Stomach pain	2	5	21	10	7	5	8	7
Scabbies	–	5	16	8	2	2	12	7
Influenza	–	–	–	–	4	12	8	8
Others				10				14
Total	31	27	42	100	29	22	50	100

Source: Author's fieldwork
Notes: [a] less than 7 days
[b] 7 to 15 days
[c] 15 or more days

Table 7.5 Adoption of treatment of sick persons in Baganbari and Kalihati, 1983–4 (%)

Source of treatment	*Baganbari*	*Kalihati*
Government hospital	5.9	26.3
upazila health centre	0.7	0.9
Chemist shop/pharmacy	4.2	11.9
Private doctor	9.5	47.7
Homeopath doctor	1.4	6.1
Paramedic	72.3	0.5
Trained *dai*	–	1.0
Others	5.0	5.8
Total	100.0[a]	100.0

Source: Author's fieldwork
Note: [a] Some totals are rounded

also adopted in both the areas; however, their proportions were very small. An enquiry was made to understand why people are reluctant to avail themselves of the government sources of treatment where better health facilities are supposed to be available. A wide range of reasons has been indicated for not attending government hospitals and health centres. Of these, 'no hospital or medical facility in the locality' and 'difficulty of communication/bad roads' for Baganbari, and 'lack of good treatment at government centres' and 'medicines at government centres are not effective' for Kalihati were frequent answers. Overall, the general impressions of the medical treatment available at the government hospitals or health centres are not good, and this tends to discourage people from attending.

NORMS REGARDING MARRIAGE AND FAMILY FORMATION

Fertility levels within rural society depend, among other things, on a number of behavioural aspects of which norms about marriage and family formation have a significant bearing. The general characteristics of marital status of the population in the study areas indicate that in both Baganbari and Kalihati, the proportions of people who are married is relatively high, as is the proportion of widows. The proportions of divorced and separated people are very low in comparison.

Marriage is almost universal in rural Bangladesh.[2] About 49 per cent of men and 68 per cent of women are married by the time they reach the age of 30 in Baganbari; the figures for Kalihati are 52 per cent and 74 per cent, respectively. The median age of first marriage is 19 years for males in both Baganbari and Kalihati and 12 years and 14 for females in Baganbari and Kalihati, respectively.

Marriage is a vital social event when women change from the status of daughter to daughter-in-law, and subsequently, most significantly, to that of mother. It also changes the role of a male family member from that of boy, with shared responsibility in the family to an adult with increased status. But the transformation for men, who usually continue to live in their original place of residence, is gradual; for women, the change is sudden as they move into a new social and spatial environment.

In the study areas, it is held that 'marriage is important to keep on family lineage – the *bangsa*. For the Hindus, the objectives of marriage are three-fold: religion (*dharma*), progeny (*prajan*) and pleasure (*rati*). It is believed by some Muslims and Hindus alike that the pairing of marital partners is spiritually predetermined. But there are others who do not believe in such predetermination. Nevertheless, such a belief, undoubtedly symbolises the importance of marriage and inhibits marriage dissolution. For both Muslims and Hindus, marriage is also a religious function coupled with social and/or legal rituals. Strictly speaking, a Hindu marriage is a sacrament, while Muslim marriage is a contract under religious sanction.

Age of marriage

Most respondents believe that boys and girls should get married at the age of 25 and 18, respectively. 'It is good for health to marry at the right time' is a common statement. However, belief and practice differ considerably. The median age of first marriage is low in both Baganbari and Kalihati and it is generally perceived that 'if the girl is not married in time (soon after puberty), her youth will be spoiled' and that 'people will speak ill of girls kept unmarried long in the family'.

Reaction to raising the age of marriage for girls by three years was investigated. Most women were unsure of the results or impact that would ensue. Between 24 and 36 per cent of women in the study areas indicated that such a change would be beneficial. The reasons are varied, yet significant, indicating a far-reaching impact if this kind of change was implemented. (Tables 7.6 and 7.7). A good proportion of respondents in both Baganbari and Kalihati felt that raising the age of marriage by three years would allow girls to gain experience in family formation and housework, which indicates the importance of family and household well-being in rural social structure.

Table 7.6 Reaction to increasing the age of marriage of girls by 3 years in Baganbari and Kalihati, 1983–4 (% of female respondents)

Reaction	*Baganbari*	*Kalihati*
Unsure of ramifications	37	35
Good	31	32
Bad	16	21
No difference	16	12

Source: Author's fieldwork

Table 7.7 Reasons for favouring an increase in age of marriage for girls, Baganbari and Kalihati, 1983–4

Reason	*Baganbari*	*Kalihati*
Help with child raising	24	22
Gain experience in family formation	22	21
Time to learn household tasks	21	19
Help raise the family income	16	15
Time for schooling	9	12
Allow time to know husband	4	6
Help maintain good health	4	3
Beneficial for health and child bearing	–	1
Total	100[a]	100

Source: Author's fieldwork
Note: [a] Some totals are rounded

Those who think negatively of raising the age of marriage for girls provided a wide range of reasons. Most were concerned with the attitude of neighbours and the community in general. Some, mostly those in lower land-holding categories, thought it sinful to keep eligible, unmarried girls at home (Table 7.8). These social prejudices are related to the degree of purdah observed in the rural areas and are quite consistent with the rural social structure of the country.

It may also be noted that the proportional differences with respect to responses are not substantiated by region, but by land ownership

Table 7.8 Reasons for not favouring an increase in age of marriage for girls, Baganbari and Kalihati, 1983–4

Reason	*Baganbari*	*Kalihati*
Other people's negative perception	31	36
Adult girls are prone to misguidance	9	17
Not good to leave girls unmarried	15	9
Adult girls are a burden on their families	14	9
It's a sin to keep girls unmarried	3	7
Girls adjust better to in-law house if married early	4	5
Delay of marriage of one daughter may lead to delay for others	9	3
Fear of 'bad reputation'	7	7
Other reasons	8	8
Total	100	100[a]

Source: Author's fieldwork
Note: [a] Some totals are rounded

categorisation. Some of the women on medium and large farms indicated reasons that have lesser importance than similar reasons do for the small-holding groups (Elahi 1987).

According to the Bangladesh Fertility Survey (BFS) data, the age of marriage of women has been rising noticeably, although its effect on fertility is minimal. The mean age of marriage for women in 1927 was 10.9 years; by 1957 this had risen to 13.3 years (13.9 years in urban areas). It is estimated that the age of marriage rose by 2.7 years between 1962 and 1975. These findings are consistent with the range of responses regarding mean age of marriage as discussed above, and also with the information on marriage in the study areas.

Marriages for girls in rural areas are arranged by their father and other male relatives or elders. Since Islamic law sees marriage as a contract, its dissolution is possible, and in that event, financial security is ensured for the wives through *moharana* – a fixed sum of money, part of which is payable to the bride on demand and the rest payable after dissolution of the marriage. Among the households of the small landholders, the male guardians of the woman regard *moharana* as a price and often demand part of it. Among medium and large farm households, it is considered to be more prestigious not to claim the moharana and the actual payment is rarely made (Elahi 1987).

In contrast to the religious sanction regarding *moharana*, there is a wide spread custom of dowry (money, livestock, farm equipment, ornaments, luxury items) which is given to the bridegroom at the time of the wedding. The amount of dowry is fixed by the guardians of the bridegroom and there are incidents of broken engagements due to a failure to meet the dowry payment. In the study areas, the value of a dowry ranged from Tk2,000.00 for the landless households to Tk50,000 or more for those in the large landholding category. It is held that 'it costs money to raise a son' and 'my son is competent both socially and economically – one has to pay for them'. Despite its widespread practice (especially among Hindu families), requesting a dowry is against the law; however, most cases go unreported. It is also difficult to assess the extent of marriage dissolution as a result of non-payment of dowry. Since most marriages are not registered, there exists a wide gap between what is laid down in law and what is practised. Further, most women are ignorant of the laws anyway. For example, in the case of divorce, the legal sanctions protecting women's interests are rarely upheld.

The practice of polygamy is looked down upon in the study areas, both socially and economically. In both Baganbari and Kalihati, the proportion of polygamous marriage was found to be very low (1.2 per cent).

PATTERNS OF MORTALITY AND HEALTH

During 1983–4, 136 males and 93 females were reported ill in the study households. However, it seems that illness among females is under-estimated and under-reported to some extent. In general, young children and older adults have a higher proportion of illness. However, for women, a substantial amount of illness is associated with child birth, especially during the late reproductive age period.

Decisions regarding treatment of ill people are crucial and reflect the level of patriarchal dominance prevailing in the rural society. Almost as a rule, the decision regarding treatment is taken by men. They account for over 54 per cent of all decision making regarding the treatment of the sick. Most of the sick were treated by a private doctor in Kalihati, and by a paramedic in Baganbari. The proportions of those who sought treatment in government health centres/hospitals were low (see Table 7.3). The underlying proposition of seeking treatment has always been the impact of decisions taken at the household level and whether the patient is male or female. It is evident that males get better medical attention than their female counterparts, which is partially explained by differential perceptions of sickness and mortality among boys and girls.

Differential mortality: perception of wives and husbands

The respondents were asked about their perception of mortality among boys and girls. Reasons for differential perceptions reflect beliefs about health, food, nutrition, and the availability of facilities. Those who think that infant and child mortality is increasing, blame inadequate health infrastructure followed by insufficient medicines and doctors. In contrast, those believing that child and infant mortality is declining have indicated the opposite: that there are improved health facilities and availability of treatment. Those responding with neutral answers continue to suffer the same problems as in the past, such as insufficient food, and have no personal experience to suggest that the child and infant mortality rate is changing.

In-depth interviews revealed that most people are apathetic towards government medical and health facilities at the local level. It is also generally believed that 'one cannot expect good treatment in the *upazila* Health Centres', that 'they give not good medicines' and that 'doctors demand money for treatment which should be free'. The members of most medium and large land-holding farm households usually attend private clinics in nearby towns: 'at least you get good medicines from the private doctors' (see Elahi 1987).

Perception of sex differentials in infant and child mortality reveals certain significant facts. In a patriarchal social set up, where male children are valued most and where it is recognised that they are biologically weaker in infancy, the responses reflected an obvious parental concern regarding the survival of male children. Most respondents, particularly wives, thought that boys are more likely to die before they reach adulthood. Those who thought that girls are more prone to sickness and death in infancy suggested that girls are disease prone and of weaker constitutions (32 to 45 per cent) and suffer from a lack of care and medical treatment. On the other hand, those who held that boys die most blamed high mortality and disease proneness among boys. These reasons, however, are less realistic since boys are valued more than girls in rural society and extreme care is taken of them in infancy. Women, in particular, are found to regard sons as the prime sources of old-age support. Further, despite the responses that support a lack of discrimination in mortality of boys and girls, information and records from the *upazila* health centres and government dispensaries in the study areas indicate that boys were given medical attention at an earlier stage of sickness than the girls. A higher proportion of male children were brought for treatment than the girls in these centres. The trend is more prevalent in the households of small land owners and in those having more than one girl in the family. The perceptions expressed as to the mortality of boys reflect the familial and social concern about the possible survival of the male children and perhaps overshadow the respondents' true feelings about child and infant mortality. On the other hand, some respondents indicated that, 'yes, people neglect daughters' treatment when they are sick' and 'daughters are not taken care of well if they are too many'. Such attitudes are reflected in the perceptions of child mortality in both areas and are found to be similar in other areas of South Asia (see Khan *et al.* 1985).

HEALTH NORMS

Related to the condition of mortality is a number of practices observed by rural women and men in birth delivery and illness. These norms directly affect mortality levels and indirectly affect a range of phenomena which influence human activities.

Birth delivery practices

In most households, the delivery of babies is assisted by the relatives of the mother, followed by *dai* (traditional birth attendants). A small proportion of women reported having had a trained nurse/*dai* and doctor to assist in the birth. But cases of unattended births or births attended solely by a female elder in the family are not rare. Those who depend on doctors and trained *dai* for delivery are mostly members of the large landholding groups. In other cases, apart from unskilled *dai*, older women, who are either relatives or neighbours, played an important role. Since the majority of local doctors are male, it is unusual for them to be called in to the birth unless there is some complication. Otherwise, local *dai* and relatives and/or older relatives are relied upon.

As has been seen earlier, the extent of generally negative attitudes regarding government health facilities and the alleged demand for money by the health personnel discourage many from seeking to deliver their babies at these facilities. The local *dai*, who have been visiting the mother during her pregnancy, ask for less money for assisting in the delivery. Very often they are paid in both cash and kind, the latter usually comprising one or two saris and some rice. The total value paid for assisting in the birth ranges from Tk80.00 to Tk300.00. If the baby is a boy, the remuneration may increase to Tk500.00. Thus, not only is there a general scepticism of the local government-run health facilities, but on economic grounds, most families prefer the services of a local *dai*.

Women prefer their babies to be delivered at home, but have a preference for their parents' home in delivering the first born. Mothers also insist on bringing their pregnant daughters home for the first birth in order to ensure adequate care for both the daughter and infant. In most families, a separate room is prepared for the birth. If there is a shortage of space, as in the case of the landless or small landholders, the living room is divided with a makeshift partition. Among the Hindus, it is common to erect a purpose-built shed which is dismantled and burned afterwards.

Pre- and post-natal care for women is scant at the formal level. Most women from small landholding households do not engage in courses organised by the government health units. Some of the educated and well-to-do women visit the government hospitals or private doctors at least two to three times in order to seek advice relating to pregnancy and child care. In most cases, the main sources of such advic elder family members, i.e. mother, mother-in-law and elder sis sister-in-law.

Illness and health-care practices

It was also one of the objectives of this study to look into the extent of differential health-care and feeding practices for selected life stages of males and females in the study areas. There are no apparent differences between male and female children in breast feeding and administering liquid foods (cow or goats milk, barley and rice water). However, there is evidence to suggest that discrimination does occur once children begin taking solid food. In the study area, children are introduced to solid food between the ages of 6 and 10 months starting with mashed rice with fish or egg or vegetables. A male child gets one egg almost every day if the parents can afford it; this is not offered to a baby girl. The discrimination is more visible at the ages of 3 or 4 years. The male children eat with adult males while female children eat with mother and other female household members, all of whom tend to eat what food is left after the men finish their meals. It is still held, particularly in the rural areas, that 'men need more food – they do heavy work and work in the fields'. In some respects this may be true, but is certainly a false assumption to make of the nutritional requirements of boys and girls at a very young age.

In general, parents prefer their children to be treated by a local homoeopath doctor as they believe that homoeopathic medicines are more suitable for babies. However, here too a significant discrimination against female children can be observed. Overall, a girl receives appropriate medical attention if she is the only child or only daughter among many sons. Otherwise, at the initial stage of illness, she is ignored and her illness is rationalised as either being trivial or a characteristic of her general level of health. It was also observed that many mothers wanted to give appropriate medical attention to their daughters but that their economic situation (as in landlessness), their low status in the household (as a daughter-in-law), having too many female children or being ignorant of the medical facilities available, all act as hindrances to proper decision making. Often, different sex differentials prevail due to the mechanisms in force regarding family-level decision making which is predominantly made by male members of the household, most often the husband.

In a way, the overall mortality and health situation discussed above is related to the total social and economic value of male children in the rural areas. This aspect, as has been observed in rural areas, greatly affects the preference for family formation in terms of desire for male children by the respondents.

CONCLUSION

...t sex discrimination in favour of males operates at ... manifested in the norms of family formation, birth ... health practices. This pattern continues throughout ...s and negatively affects women. The relevant norms ...ough to influence perceptions about the degree of ... and female children in the society, and about family formation, ... support, economic usefulness of children and differential mortality rates. There is a great deal of difference in gender discrimination on these counts according to the developmental level of each of the two study areas.

The level of fertility is influenced substantially by a number of behavioural conditions of which norms about marriage, age at marriage, old-age support and economic usefulness of children (mainly male children) and family formation have a significant bearing on rural society.

The proportion of persons falling ill and seeking treatment are biased towards males in both study areas. It seems that sickness among females is under-reported and under-estimated to some extent. The decision taken regarding treatment of the ill is crucial for a number of reasons and reflects the level of patriarchal dominance prevailing in Bangladeshi rural society. Almost as a rule, decisions regarding medical attention are taken by the husbands/fathers who are usually the household heads and chief earners of household income. Whatever the range of medical facilities available, access to treatment has always been affected by the decision that has been adopted at the household level and whether the patient is a male or a female household member. In the case of illness, males tend to get better attention, which reflects the differential perception regarding sickness and mortality in male and female children.

It has been observed that traditionalism affecting the status of women and the recognition of their activities both within the family and at the community level is deep rooted and influences the socioeconomic norms and value systems influencing men. These, in turn, significantly affect the fertility, mortality and health norms in rural Bangladesh.

8

SEASONALITY, WAGE LABOUR AND WOMEN'S CONTRIBUTION TO HOUSEHOLD INCOME IN WESTERN INDIA

Elizabeth Oughton

INTRODUCTION

This chapter considers the significance of the contribution of women's earnings to the income of the household. It is based upon field work in Sangli District, Maharashtra. In particular, it is concerned with the seasonality of income flows and their importance at times of seasonal stress.[1] Households in predominantly rural areas devise a number of strategies to cope with the regular pattern of seasonality of production and employment. Poorer households, however, may find even the expected pattern of seasonality difficult to accommodate. The nature and proximate cause of seasonal stress will therefore depend upon the social and economic characteristics of the household.

Raikes (1981) has described seasonal stress as arising from food shortages combined with periods of high labour input requirements. Farmers with small landholdings are affected when they are unable to produce sufficient to carry them from one harvest period to the next. For the landless or those with very small plots the pattern of seasonal stress will differ and be characterised by declining exchange entitlements as grain prices rise and earnings fall, due to falling seasonal labour demand and/or declining real wages. Fluctuating labour incomes are highly significant to poor households. In the first case, poorer households receive a higher proportion of their income from labour. Secondly, they are more vulnerable because they have fewer reserves, and are less likely to be able to borrow to carry themselves through emergencies (see Lipton 1983).

Jiggins (1986) has argued that the degree of flexibility of the household is important in increasing its ability to withstand seasonal stress and points to the importance of examining:

> the more dynamic calculus in which household members complement each other's contribution to livelihood stability across seasons by maintaining the capacity to transfer resources in and out of the sub-systems which together constitute their livelihood.
>
> (1986:9)

One aspect of this 'dynamic' calculus is the degree to which different members of the household move in and out of the paid labour force. Thus for example Ryan and Ghodake (1980), using data from six south Indian villages, find considerable seasonal variation in labour-market participation rates for both males and females although across the year as a whole, labour-market participation is low: 'On average over these six villages . . . 30 and 37 per cent of available male and female labour, respectively, participated in work outside their own farms, households and businesses.' (1980: 11). In the two villages surveyed in Sholapur District, the immediate neighbour of Sangli District where the present study took place, the maximum seasonal coefficient of variation in male labour-market participation was 29 per cent and of female participation 22 per cent. In two of the villages male and female employment periods coincided. In the remaining four villages, the peak and slack periods for males and females differed, mainly due to the gender-specific nature of the agricultural operations being carried out. The data presented in these studies do not permit an estimate of seasonal earnings as opposed to employment.[2]

This chapter looks at just one aspect of seasonal flexibility, that is, the contribution of women's earnings to household income. The essential argument is that female earnings from wages and salaries are important to family well-being at all times,[3] but that at periods of seasonal stress they become vital to the maintenance of the household.

THE STUDY AREA AND THE PATTERN OF SEASONALITY

The data used in this analysis were collected during field work in Maharashtra in Western India between 1982 and 1984. Panel data were collected from a stratified random sample of 94 households of which 87 remained for the full period of the study. The data were collected over 12 rounds at monthly intervals. Sangli District, in which the study was based, lies in what is officially described as a drought-prone region, the major characteristic of which is an annual rainfall of less than 750 mm. This area covers 12 districts from north to south lying in the rain shadow of the Western Ghats. Figure 8.1 shows the districts containing drought-prone talukas and the location of the study.

The climate of Maharashtra is influenced by two monsoons. The

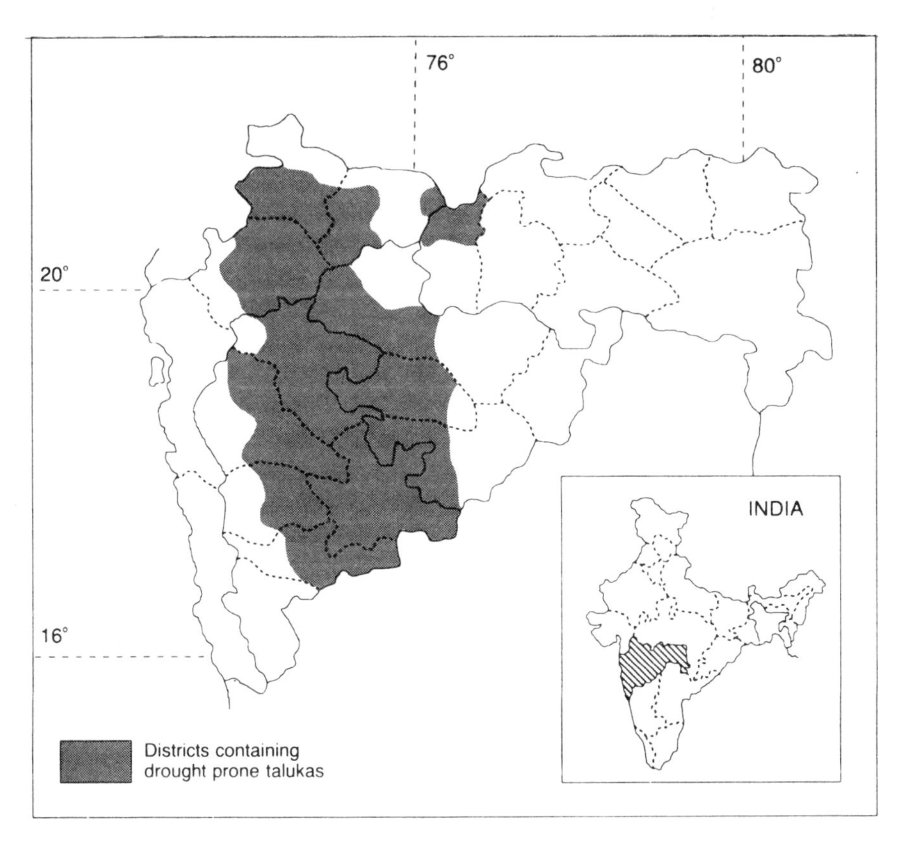

Figure 8.1 The study area and location of districts containing drought prone talukas

south-west monsoon originating over the Arabian Sea is the most important to the state as a whole. Rain from the south-west monsoon falls from June to November. The north-eastern monsoon lasts from November to January, but is much less significant in western Maharashtra. The two monsoons lead to a bimodal pattern of rainfall with peaks in June and September. Figure 8.2 shows the rainfall pattern during the study. The year can thus be divided into three seasons: the monsoon, from mid June to late September/early October; winter, from November through February; and summer, characterised by high daily temperatures and negligible rain fall from March through June.

The bimodal rainfall pattern provides two main cropping seasons: the *kharif*, in which crops are planted in June following the first rains and harvested between late October and early December, and the *rabi*, starting in September with harvesting from late December to early

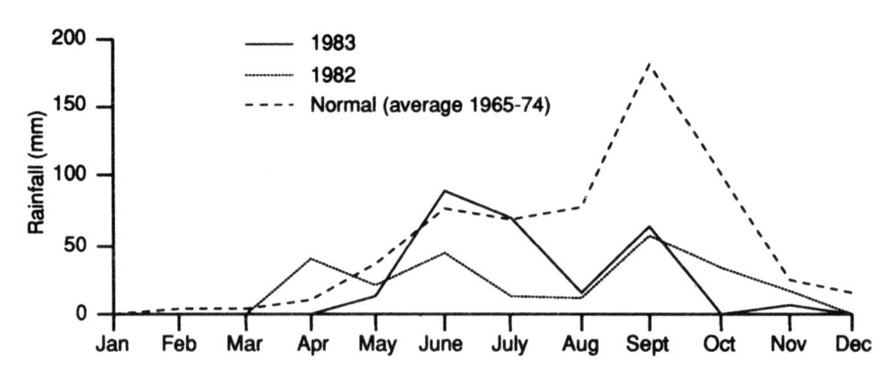

Figure 8.2 Annual distribution of rainfall, 1983–4
Source: Mamlatdar's office, Kavathe Mahankal

February. The most important food grains are *jowar* (sorghum) and *bajra* (pearl millet). The cropping patterns of the major food grains are shown in Figure 8.3.

The relative importance of each crop season depends upon the pattern of rainfall and the farmers' estimation of the risk of monsoon

Table 8.1 Area and yield of major cereals in Kavathe Mahankal block, 1979–80 to 1983–4

	Area (hectares)				
	79–80	*80–1*	*81–2*	*82–3*	*83–4*
Kharif season					
Hybrid jowar	2,300	3,782	4,650	2,250	6,390
Local jowar	6,100	7,406	7,236	500	5,100
Hybrid bajra	80	281	509	500	3,885
Local bajra	12,611	16,553	16,268	11,520	7,515
Rabi season					
Hybrid jowar	12	9	39	40	240
Local jowar	15,148	9,619	10,310	12,290	12,500
HYV wheat	3,875	844	1,560	2,617	1,250
Local wheat	230	834	625	380	250
	Yields (kg/ha)				
	79–80	*80–1*	*81–2*	*82–3*	*83–4*
Kharif season					
Jowar	1,004	941	1,001	625	770
Bajra	1,206	1,093	1,001	291	220
Rabi season					
Jowar	815	1,000	997	835	na
Wheat	686	1,088	1,030	1,069	na

Source: Sangli District Zilla Parishad

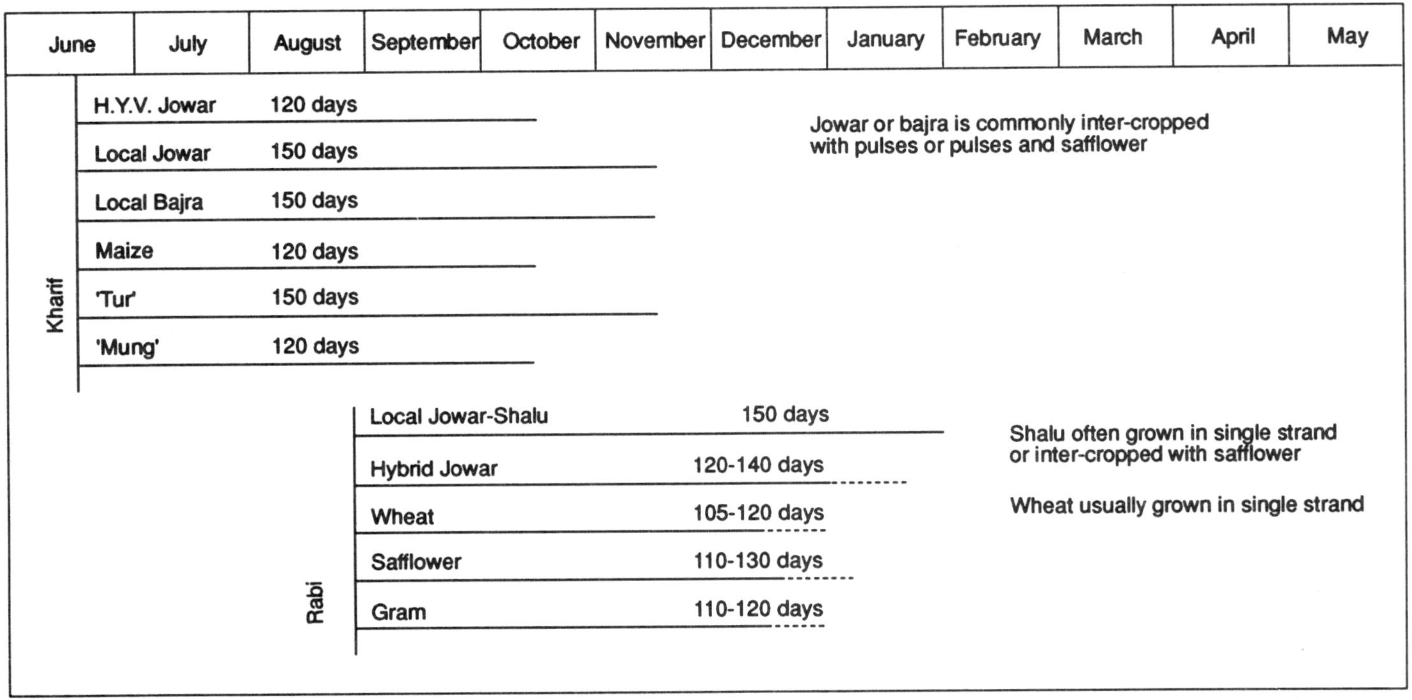

Figure 8.3 Cropping patterns common in Kavathe Mahankal taluka
Source: Agricultural adviser, Panchyat Samiti

failure. The important point to note, however, is that the pattern of seasonality arising from crop production may change from year to year. The periods of seasonal stress may vary, and they may affect different groups in different ways.

This chapter is concerned with households that hired out labour during the study period. Distribution of land ownership in the village is highly skewed; 15 per cent were landless and of those with land, 30 per cent had less than one hectare.[4] The harvest immediately preceding the beginning of my fieldwork had been very poor. Table 8.1 shows the area and yield of the major cereal crops in the *taluka* during that period.

The decline in the area of *kharif* cereals is particularly noticeable. To a limited extent, this is compensated by the increase during the *rabi* season, but in both seasons yields fell dramatically. The region had been declared a scarcity zone at the beginning of 1983 and various relief measures were in progress. The implications of this were that not only were the landless seeking work, but so were members of farm households, particularly those who had no access to irrigated land. (For this reason I have not characterised households according to whether they own land or not.) It can be seen therefore that in the year that this study was carried out, very poor harvests and very low stocks at the beginning of the year exacerbated the usual pattern of seasonality.

Information on income, employment and prices facing the household allow us to identify the periods of seasonal stress that occur through the 12-month period to March 1984. Daily wage rates for casual agricultural labour showed little variation across the year. Male agricultural wages stayed constant at Rs6 to Rs7 throughout the year with very occasional payments of Rs10 per day for very heavy manual work, such as well digging or grubbing up sugar-cane roots. Female daily agricultural wage rates remained at Rs3 per day until the end of the year when some women received Rs4.

Ryan and Ghodake (1980) found a similar disparity between male and female wage rates calculating that women received about 56 per cent of the male daily wage. This is much lower than the 80 per cent estimated by Rosenzweig (1987) on the evidence of 13 Indian states. Ryan and Ghodake also found that the lowest female wages from the three regions were paid in the drought-prone villages of Sholapur.

The average total monthly income for households which hired out female labour in the month under question is shown in Figure 8.4. Income is low, the annual average being Rs248.[5] The figures exhibit considerable monthly variation ranging from a low of Rs143 in May–June (survey round 3) to a high of Rs338 in January–February (survey round 11). However, the real value of earnings was eroded by rapidly increasing prices for food and other goods throughout the year. Table 8.2 shows two different indicators of the price increases facing these households.

Table 8.2 Price changes facing agricultural labourers, 1983–4

Consumer price index for Agricultural labour, Maharashtra (1960–1=100)			*Village survey data*	
Month	Food	Change from previous month	Month	Price of jowar (Rs/kg)
March '83	478	−9	March–April	1.30
April	489	+11	April–May	1.30
May	521	+32	May–June	1.35
June	528	+7	June–July	1.35
July	541	+13	July–Aug.	1.40
August	559	+18	Aug.–Sept.	1.60
September	572	+13	Sept.–Oct.	1.65
October	564	−8	Oct.–Nov.	1.65
November	574	+10	Nov.–Dec.	1.65
December	577	+3	Dec.–Jan.	1.60
January '84	592	+15	Jan.–Feb.	1.70
February	572	−20	Feb.–March	1.80
March	543	−29		

Source: Government of India and Fieldwork, 1983–4

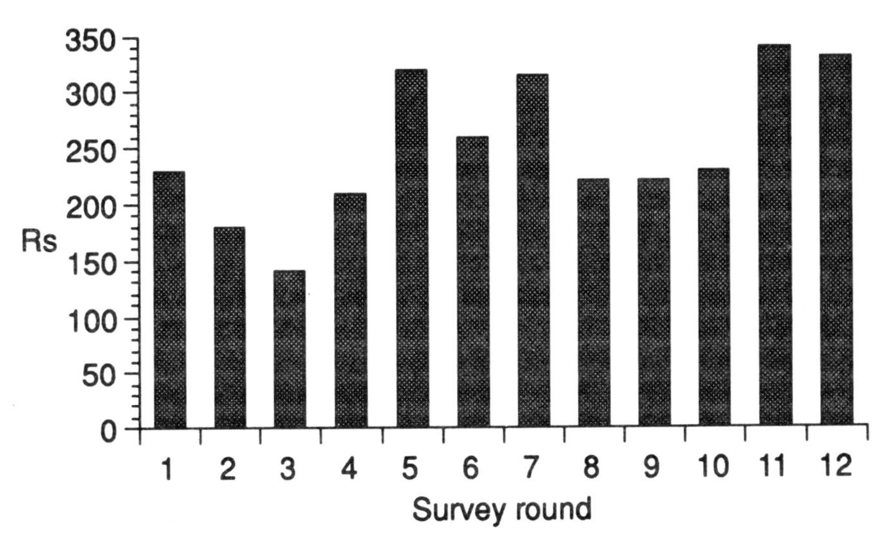

Figure 8.4 Average monthly income of households hiring out female labour

The index of prices for agricultural labourers, derived from state-level data, shows that food prices rose by 114 points, almost 24 per cent between March 1983 and January 1984. It was only following the *rabi* harvest in February that prices began to fall. Within the village, price rises were even more extreme. The cheapest price for *jowar* dropped very little following the *rabi* harvest of 1983. Prices rose throughout the year falling only 5 paise per kilogram after the *kharif* harvest. By February 1984 *jowar* prices were 42 per cent higher than they had been 12 months previously.

The picture therefore is one of increasing prices and falling real incomes all across the year, with significant troughs in survey rounds 2, 3 and 4 and 8, 9 and 10. Those households with productive land were able to harvest their own crops from November to December, but for the landless the pressure from increasing prices continued until February and the arrival of the *rabi* crops onto the market.

A pattern begins to emerge of significant stress during the first rounds of the survey. For labouring households, and especially those with few grain stocks of their own, unseasonably high prices and limited earning opportunities created a difficult position. Good monsoon rain fall through June and July ensured reasonable opportunities for work for both men and women from irrigating, weeding and tending crops, although the pressure from rising prices remained.

SEASONALITY AND EARNINGS FROM EMPLOYMENT

The labour earnings per household per month will depend on the number of people working, the number of days that they worked and

Table 8.3 Households hiring out labour on a seasonal basis

Month	*Male labourers only %*	*Female labourers*[a] *%*
March–April	52	8
April–May	60	11
May–June	60	10
June–July	52	9
July–August	60	22
August–September	60	23
September–October	58	17
October–November	58	15
November–December	56	18
December–January	53	13
January–February	54	15
February–March	51	13

Source: Fieldwork, 1983–4

Note: [a] These households do not necessarily hire out female labour only. Some may hire out both men and women, but in all cases at least one woman has been hired out during the month.

the daily wage rate.[6] Table 8.3 depicts the number of households, on a monthly basis, in which at least one member worked outside the household for a cash wage, either on a daily or weekly basis, or for a monthly salary.

Across the year there is little variation in the proportion of households hiring out male labour. The figures are slightly higher for the first half of the year than the second half, ranging from a maximum of 60 per cent in mid April to mid June and July to mid September, and a minimum of 51 per cent in February to March. The households hiring out female labour show a more dramatic pattern. In the first 4 months, the maximum is only 11 per cent and the minimum is 8 per cent of all households. As we shall see, this configuration changes when kind earnings are taken into account, although the figure always remains low in May through until mid July. The work available at the beginning of the survey was mainly land preparation and land improvement, predominantly male work. During June and July the *kharif* crops are sown. There is a slight dip in all hiring out as those households with land begin the planting of their own crops. The percentage of households hiring out female labour increases sharply in the next two survey rounds as women enter the agricultural labour market primarily to weed and tend crops. Although there was more work available, this was also a time of pressure upon the household as the prices of food grains were increasing rapidly. Between the end of May and the end of July, the price of second-quality *jowar* had increased by 19 per cent. From mid September until the end of the year, the proportion of households hiring out women then remained fairly steady at between 13 and 18 per cent.

By looking more closely at the households which hire out women, it is possible to calculate the contribution that women make to total household income. Figure 8.5 is based upon the actual number of women working in each household and their cash income.[7] For the year as a whole, females' cash earnings contributed slightly less than 40 per cent to total household income. Thus the first point to make is that the contribution of women's wages to household income is significant across the whole year. Second, Figure 8.5 shows the seasonal importance of the women's earnings. In May–June (survey round 3), female earnings account for 59 per cent and in June–July (survey round 4), they account for 51 per cent of total household income in those households that hire out female labour. Third, by referring to Figures 8.4 and 8.5 together, it can be seen that there is a correlation between low total income and a high seasonal contribution of women. Conversely, in January–February (survey round 11), when the women's proportional contribution to household income is lowest, the absolute level of total income is the highest in the year. What we cannot tell from these data alone is the direction of causality. Are women in later rounds working less because

Figure 8.5 Female contribution to household income

the household has a higher income, or are incomes low in households where women have a greater share of the work because women are paid at a much lower rate than men? There is a strong preference in Indian rural households to avoid hiring out female labour for reasons of social status.[8] Thus, the presence of women in the labour market in the May to July period in particular convincingly suggests that women were working because household income was so low that these households were under considerable seasonal stress and the woman's income was vital to the welfare of the household. These households accounted respectively for 9 and 11 per cent of those sampled, a small but significant number.

The relatively higher incomes in the later rounds do not contradict this argument. In January and February, women's income contributes a smaller proportion to total household income, even though they account for almost two-thirds of income from employment. Total income receipts are high at this time because of sales of cash crops following the harvest. This in itself is a sign of pressure as households are forced to sell cash crops when prices are at their lowest in order to meet their cash needs.

If we consider income from employment alone rather than total income, the average contribution for women workers is 61 per cent across the year and reaches a peak of 79 per cent in June–July (survey round 4). Monthly earnings are higher for females than for males despite the fact that male wages are twice that paid to women. This implies that women spent more than twice as much time in wage labour activities than did men.

Income from employment shows a distinct bimodal pattern (Figure 8.6). Immediately following the *rabi* harvest employment income is high, falls significantly in April–May (survey round 2) and then climbs steadily over the next 5 months to September–October (survey round 7). This is followed by a second drop and steady recovery to the end of the year at the next *rabi* harvest. Average female earnings do not show such obvious seasonal patterns, although there is significant monthly variation. For most of the year, male and female earnings move in the same direction; in two survey rounds there is a clear inverse relationship between the two. Again, the month of particular interest is mid June to mid July (survey round 4). The average earnings for men drop to an annual low in that month, but it is more than compensated by the increase in average female earnings. The obvious explanation would be that men are busy with planting on the home farm and women are earning cash to support the household. This would also seem to be a satisfactory explanation of the January–February figures. It is the beginning of the *rabi* harvest, a period of abundant work, especially if the harvest is good. Post-harvest work, particularly threshing, is predominantly carried out by women. Men may be harvesting on their own farms while women find employment in a relatively buoyant labour market. Thus the data suggest that there is a complementarity of livelihoods between men and women in the household, both contributing to family security.

This is, however, further evidence that this contribution to seasonal security from female household members represents an additional burden for women. In a separate study of these data looking at labour use on own farms, Raghuram (this volume, Chapter 9) found that in all but two months of the year: March–April (survey round 1) and February–March (survey round 12), female labour time exceeded that of men.[9] This was particularly significant through mid April to mid July when the average time spent by women on agricultural work was over double that spent by men. It would therefore appear that home farm and agricultural labouring work are not straightforward alternatives but represent an additional burden on the women of the household.

SEASONALITY AND PAYMENTS IN KIND

As expected payments in kind demonstrate a distinct pattern of seasonality and are closely associated with post-harvest and thus female activity. Table 8.4 shows the pattern of kind payments on a household basis, i.e.

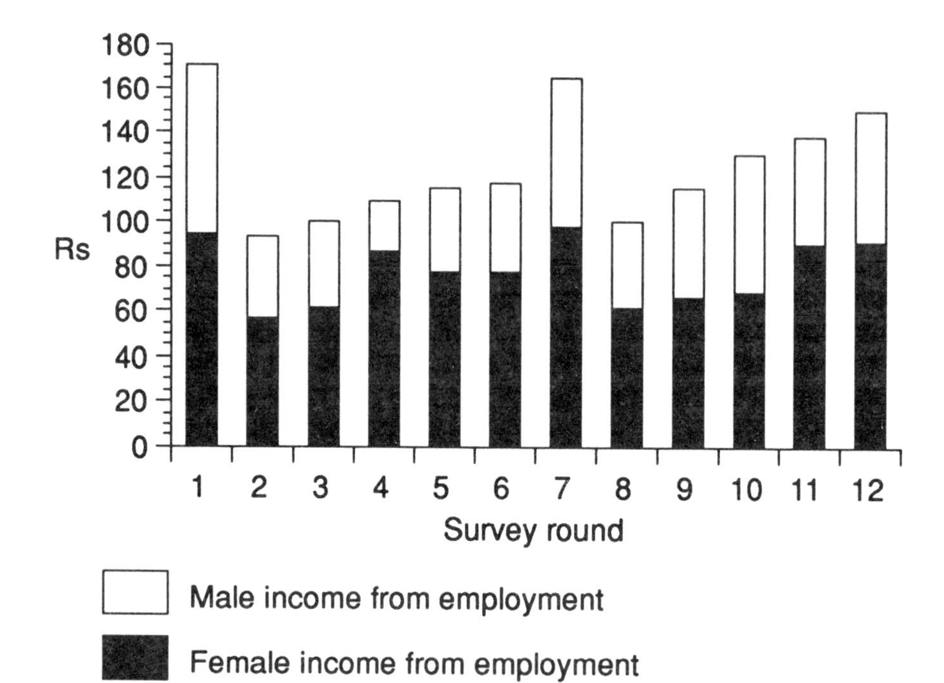

Figure 8.6 Contribution to household employment income

Table 8.4 Number of households receiving payments in kind

Survey round	*1*	*2*	*3*	*4*	*5*	*6*	*7*	*8*	*9*	*10*	*11*	*12*
Male labour	1	1	0	0	0	1	1	0	0	0	0	0
Female labour	20	1	1	0	1	1	7	12	11	4	2	10
Don't know	3	1	0	0	0	0	0	2	0	0	0	2
Total	24	3	1	0	1	2	8	14	11	4	2	12
% of sample	28	3	1	–	1	2	3	16	13	5	2	14

Source: Fieldwork, 1983–4

those households that have received at least one kind payment during the month are recorded.

Only four households during the year received kind payments for work completed by men.[10] Concentrating then upon female workers, Figure 8.7 shows the seasonal pattern in the proportion of households hiring out female labour for cash payments alone, cash and kind

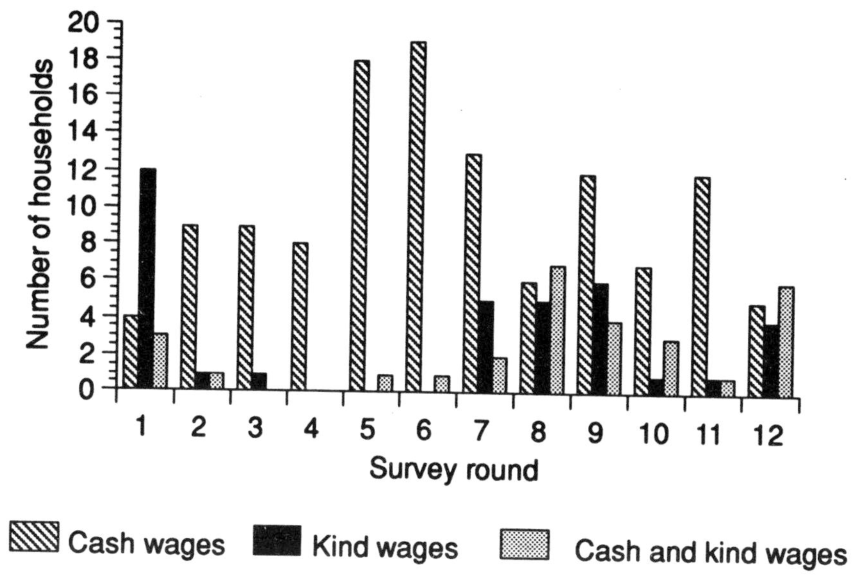

Figure 8.7 Households hiring out women for wages paid in cash, kind or both

payments and kind payments alone. Once kind payments are included, the percentage of households with female labourers rises considerably. For six months of the year more than 20 per cent have at least one woman employed outside the household. The number of households receiving kind payments increases dramatically in the two main post-harvest periods, October–December (survey rounds 8 and 9) and February to mid April (survey rounds 1 and 12). There is however also a significant amount of payment in kind immediately prior to the harvest (survey rounds 7 and 8). This was grain from the previous year's harvest that farmers were clearing from stores. Much of this 'old grain' is *bajra* rather than *jowar*, a grain widely eaten in this area but usually chosen for variety rather than as a daily staple. This raises two questions: what was the value of the kind wage compared to the cash wage and does this have any implications for the study of women's seasonal contribution to household food security?

Examination of data does not show any systematic pattern which would allow us to answer the first of these questions. The most common rate of payment in kind at the beginning of the year was two kilograms of *jowar* per day. Assuming a retail price of Rs1.30 per kilogram, the value of kind wage, Rs2.60 was less than the cash wage of Rs3 per day. This argument, however, depends on the quality of the *jowar* paid in kind and the preference of the household for the quality of *jowar* that it purchases. First-grade *jowar* costs between Rs1.65 and Rs1.75. If the

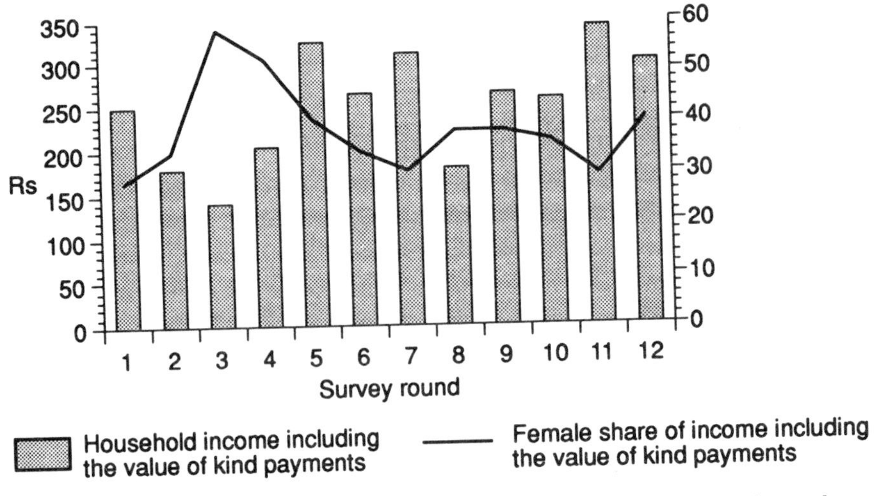

Figure 8.8 Household income showing the contribution of females and including the value of kind payments

kind wage was of first-quality *jowar* and this was the type of *jowar* that the household would choose to buy then the value of the kind wage was higher than the cash. Later on in the year there is evidence of kind payments varying from 2 to 4 kilograms per day, but without further information on the quality of the grain and the work being carried out it is not possible to determine whether the cash or the kind wage was of higher value.

It is possible to determine from the data the total quantity and type of grain received by each household.[11] Figure 8.8 depicts the contribution of female earnings to total household income on a monthly basis, including an imputed value for kind earnings, and Figure 8.9 shows the contribution of female earnings to total income when households are recorded and grouped into those with cash earnings only, kind earnings only and cash and kind together. When kind payments are taken into account the significance of female earnings decreases very slightly from 39 per cent to 38 per cent for the year as a whole. This is because, in general, these households have higher incomes. The picture and the significance of the women's role now become much more complex. If we consider first households receiving kind payments alone the process of averaging disguises an important feature of kind payments. Figure 8.9 shows that kind payments are most critical in May–June (survey round 3) and October–November (survey round 8). These are months of high seasonal stress. May–June is the hot season immediately prior to planting and October–November the month immediately preceding the

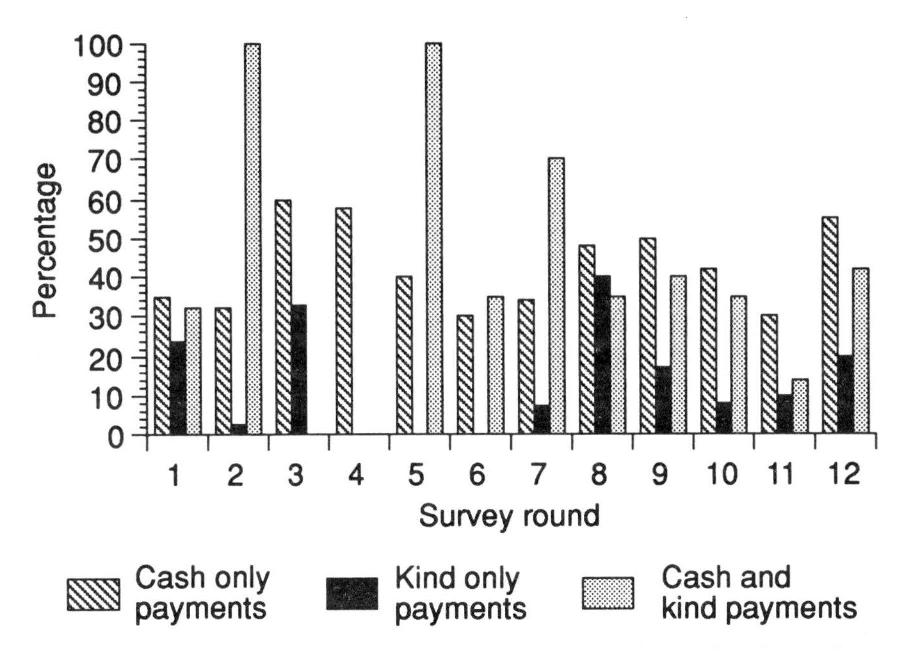

Figure 8.9 Share of total household income accounted for by female earnings of different types

harvest. The total average monthly income for the households receiving these payments is Rs156 and Rs138.70, respectively, the two lowest figures for the year. Survey round 8 is particularly interesting since this group consists of five households for whom an average of more than 40 per cent of their very low income is coming from kind payments. To these households, payments in kind to female workers are highly significant. Apart from these households kind payments alone are not of major importance.

CONCLUSION

We are now in a position to review both the annual and seasonal contribution of women workers to household income. Firstly, as Figure 8.9 shows, there is a significant difference between those households that receive only kind wages and those that receive cash or cash and kind payments. Across the 12 months surveyed, the latter groups accounted for between 9 per cent and 24 per cent of the sample. Secondly, for those households in which women worked for cash wages (this will include those who receive some payments in kind as well) in every month the value of the cash earnings of women exceeds the value of male earnings. It must be remembered that given the prevailing wage rates, women needed to work twice as long to achieve the same level of earnings as men. Thirdly, the data show that women's contribution increased during periods of seasonal stress. Fourthly, the study of the households earning cash income suggested that there was some complementarity in the use of the labour resources of the household. Off-farm work is an additional burden to these women, not an alternative to their domestic reproductive or productive labour.

With respect to the households in which women worked for kind wages only, it was shown that in two very difficult months, these earnings were vitally important. For the rest of the year, even in the immediate post-harvest periods, the importance of kind relative to cash earnings was low. It is hypothesised that hiring out labour for kind payments does not have the same social stigma as hiring out female labour for cash. Women from 'better-off' households may work on a neighbour's farm at harvest time, not strictly as exchange labour, but receiving payments for mutually co-operative work.

The factors that affect the entry of women into the paid labour force are low income, the gender-specific nature of tasks, alternative employment for male members of the household on their own land and social pressure, especially in higher-status households. At times of seasonal pressure, women are coerced into the workforce despite their numerous household responsibilities.

9

INVISIBLE FEMALE AGRICULTURAL LABOUR IN INDIA

Parvati Raghuram

INTRODUCTION

Indian agriculture is purported to be a male dominated activity. The 1977–8 National Sample Survey data estimates the female rural labour force as 61 million as against 139 million males. Seventy per cent of the total persondays is accounted for by males and about four-fifths of the total workdays are spent in agriculture. Thus official data confirms Boserup's (1970) analysis of the Indian farming system as male dominated, a characteristic of plough cultivation. Women are either invisible or considered as agricultural helpers. There is a need to reassess these conclusions and to bring out inherent biases in the definition of workers and in data-collection techniques.

The Indian government has changed its definition of 'workers' in every census. This has led to problems of non-comparability of data. There have also been inherent biases in the classification of 'workers'. For example, the 1961 census asked a loaded question based on interpretation of key words such as 'cultivator', 'agricultural labourer', and 'household industries': 'Are you working as cultivator, agricultural labourer, working at household industries or working under any other category other than the three mentioned?' The 1971 census asked: 'What is your main activity? What is your other activity?' (Government of India 1971). In Indian society where the social ideal prescribes the withdrawal of women from the workforce, the women usually see their main activity as housekeeping and child care. Thus female contribution to agriculture remains unenumerated. In the 1981 census, respondents were questioned about their main activity in the previous year and about any other work they may have done in that year. The respondents were then categorised as main workers if they had worked more than 183 days in the year, and marginal workers if they had worked less (Government of India 1981). The use of a minimum time criterion has also led to the exclusion of many female labourers whose contribution to field labour is concentrated into a few months of peak labour demand based on the seasonal requirements of the local crop calendar. The practice of interviewing the household head, who is generally assumed to be a male, has compounded the problems of fluctuation and under-enumeration of female participation in agricultural labour (Anker 1983).

The analysis of time-use data goes a long way in correcting these imbalances. However, methods of data analysis have also served to perpetuate the male bias in enumeration of agricultural contributions. Some empirical studies dealing with total 'farm labour' and using time data have, however, equated women's work time with half or three-quarters that of a male operator, on the assumption of standardising the differences in wage rates (Kahlon *et al.* 1973) and/or productivity. This assumes a parity between wage rates and productivity; productivity being standardised for all categories of men, irrespective of age, state of health or nutritional status. A critique of such praxis has been slow to evolve and has been spearheaded by feminist economists (Sen 1988) but has failed to change current practices significantly. Such 'standardisation' still forms the basis for articles in such leading Indian journals as *Studies in the Economics of Farm Management*. There is a need for a sounder empirical base in order to place the contribution of women to Indian agriculture in its correct perspective. Hopefully, this chapter goes some way towards achieving these goals.

This chapter is based on data collected over 12 months in 1983–4. The present analysis is confined to the contribution of adult family and wage labour in work on 76 farms in Karoli, Sangli district, Maharashtra. The amount of total agricultural work done by women follows the seasonal pattern of the crop calendar. The total agricultural work is disaggregated in terms of number of persons and the time spent by various categories of labour in agricultural activity. The chapter concentrates upon labour use on own farms and is not concerned with that which is hired out.

SIZE OF ADULT FAMILY LABOUR FORCE

In our study, the total number of female householders engaged in agricultural activity exceeds that of male householders. This pattern is observed over 7 months of the year (Figure 9.1). There is a seasonal peak of female labour in October–November. Female contribution to familial labour allocation of agricultural tasks is maximum during the monsoon and in winter, with two seasonal peaks in September–October–November and February–March. The dry season is the slack period, characterised by greater male contribution. The gender specificity of tasks is such that male labour is used more evenly across the year. The

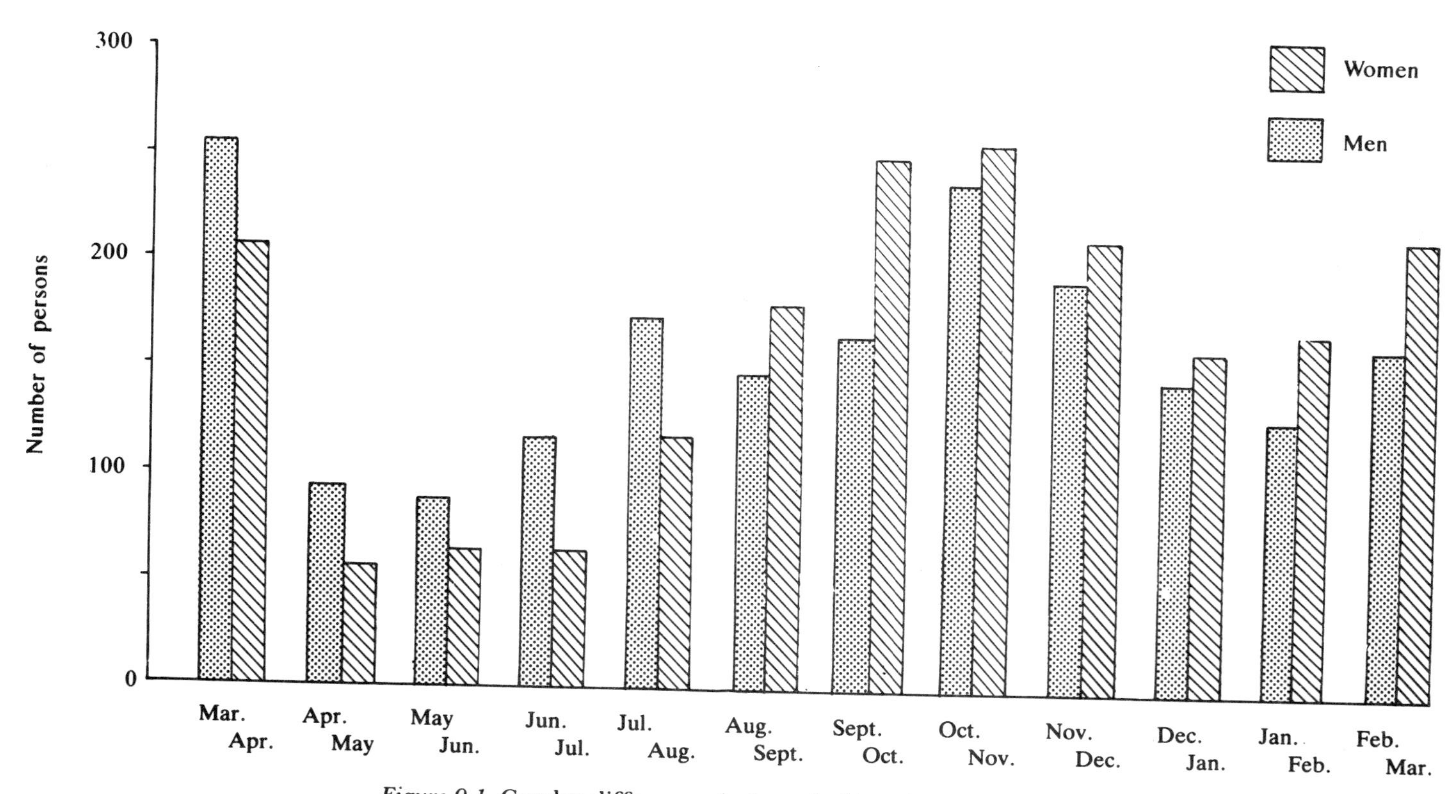

Figure 9.1 Gender differences in household agricultural labour

significance of female family labour in the agricultural year then, is clearly established.

TIME ALLOCATION ON OWN-FARM WORK AND SEX TYPING OF TASKS

The data for time spent on specific tasks was collected by interviewing and using a one-month recall. The working day has been divided into three time units. Time unit 1 indicates 0–4 hours spent in one agricultural task on one crop or on animal care. This is equated with about half a day's work. Time unit 2, or a full day's work extends for 7 hours. These figures have been aggregated for the number of days spent on that operation. In cases where the respondent reported not knowing how long the work was done for, or if the number of days spent on a particular task is unknown, the data has been adjusted to exclude such individuals from the analysis.

Measures of time spent in various agricultural operations indicate that even in the slack period, women dominate the agricultural scene. The only large positive differential for male labour is in March–April, the peak of the ploughing season (Figure 9.2). The maximum number of hours spent by women on agriculture on their own farms is in February–March when 3,292 time units of work were put in. The minimum was in May–June when 1,852 time units were counted. Thus sex typing of tasks and seasonal variations in labour demand determine the work patterns of labour. Field labour is dominated by men at most times of the year while animal care is a female task. In the field, women provide the major labour force for activities such as harvesting–threshing and sowing–planting–weeding–pruning. Ploughing–hoeing–harrowing, application of fertilisers, pesticides or irrigation water, storing and packing food and land-improvement activities, such as building bunds and deepening wells are male tasks. Management and supervision are best described as family occupations with male domination being replaced by female household labour during periods of peak demand.

Sex specificity of tasks is more rigid for women than men (Figures 9.3 and 9.4). Men often participate in activities categorised as female but the reverse is rarely observed. There is some evidence that men increase their contribution to animal care in order to free women for female-specific agricultural labour, such as harvesting and threshing. There are some instances of women participating in ploughing–hoeing but its occurrence is extremely rare. However, women do participate in heavy field labour, in land-improvement work and in application of modern farming techniques including irrigation and fertiliser application. Interchangeability of tasks is not very common and is not an indicator of complementarity in labour allocation between the sexes. Women's job

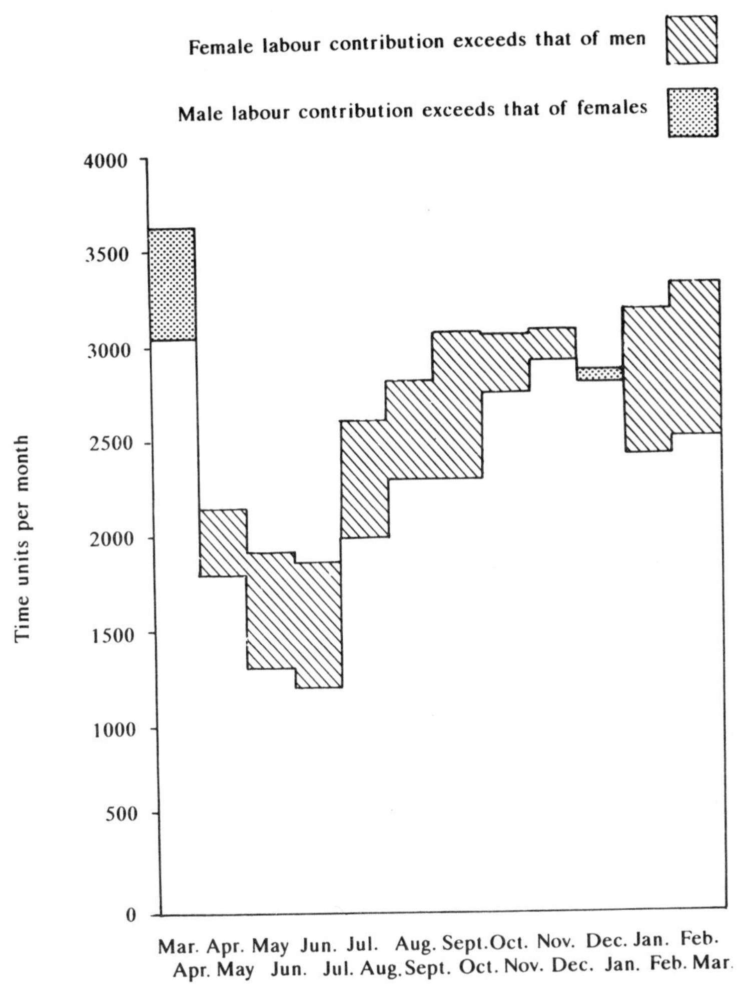

Figure 9.2 Gender differences in time spent on agricultural labour

alternatives remain limited, and interchangeability of tasks usually described as gender specific only takes place between adult members of the household for work on their own land. Even for similar tasks there is no parity between male and female wages. The social norms informing the gender specificity of tasks is of greater significance in defining wage labour than household labour and among children than among adult household members.

The average time spent on agricultural activities by an individual in one month, or the average length of the 'workmonth' is greater for women than for men. Even national statistics acknowledge that the agriculturally

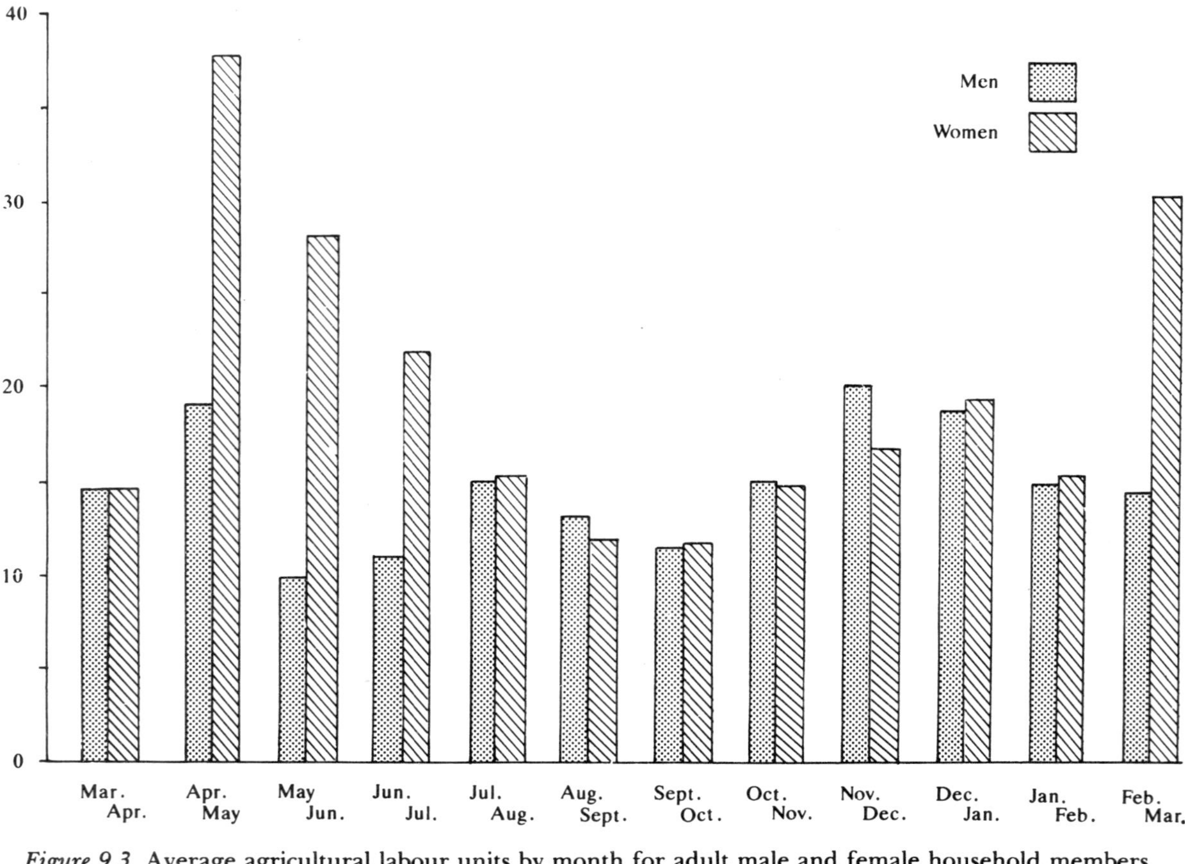

Figure 9.3 Average agricultural labour units by month for adult male and female household members

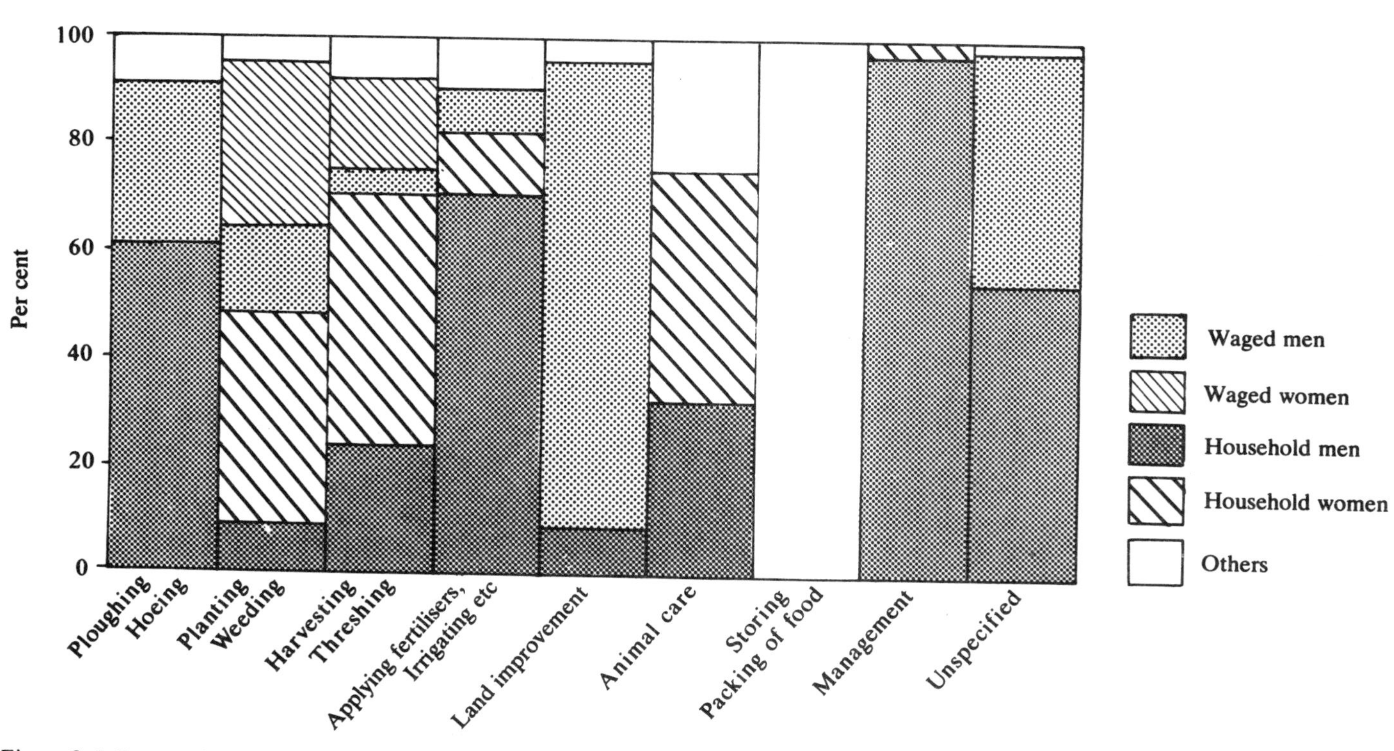

Figure 9.4 Proportion of total agricultural labour done by adult men and women – household and wage, peak period: Nov.–Dec.

productive activity of a farm woman occupies 83 per cent of her total workday as compared to 78 per cent for men. Our data also reveal entirely different patterns of seasonality for men and women. Contrary to expectations, the worktime for women on their own farms is greater during the slack period. There are two possible reasons for this (Figure 9.5). Firstly, the family responds to peak labour requirements in the harvesting–post-harvesting period by increasing the number of persons employed. Hence, the individual load of own-farm work is reduced during the peak period. Secondly, the evidence is also skewed by the presence of a large number of female peasants who take up agricultural wage work on larger landholdings during periods of peak female labour demand, reducing the time spent working on their own farm. The dry season is also the period of fodder scarcity, necessitating increased time-use on fodder collection, a task usually done by women and children. The longest working hours are put in by wage labourers, but their incidence is low in comparison to familial labour.

LABOUR ALLOCATION IN DIFFERENT LANDHOLDING SIZES

Women's contribution to agriculture is not homogeneous in society. We need to see the linkages between gender-based differences in agriculture and characteristics of the individual and the household. Caste, class and religion are some of the significant determinants of intra-regional disparities in familial labour allocation. The dominant social norms subscribed to the withdrawal of women from the agricultural labour force with increasing wealth. We have identified similar processes operating in households with larger landholdings.

The households included in this analysis possessed some land of their own towards which some family labour input was made. The time spent on animal care has been subtracted for the present since the size of the landholding was not seen to have any direct bearing on time spent in animal care. The households have been divided into five categories:

1 smallest peasant holdings 1 hectare or less;
2 households having between 1 and 2 hectares of land;
3 households with between 2 and 4 hectares of land;
4 households with between 4 and 6 hectares of land;
5 largest landholdings of at least 6 hectares of land.[1]

The time spent in agricultural activity and the size of the labour force in the above five categories, standardised per hectare, reveals a decrease in the contribution of female labour as landholdings increase in size. Similar patterns have been observed in the Andean region (Deere 1982) and in Java (Hart 1980). Gender specificity of tasks is of greater

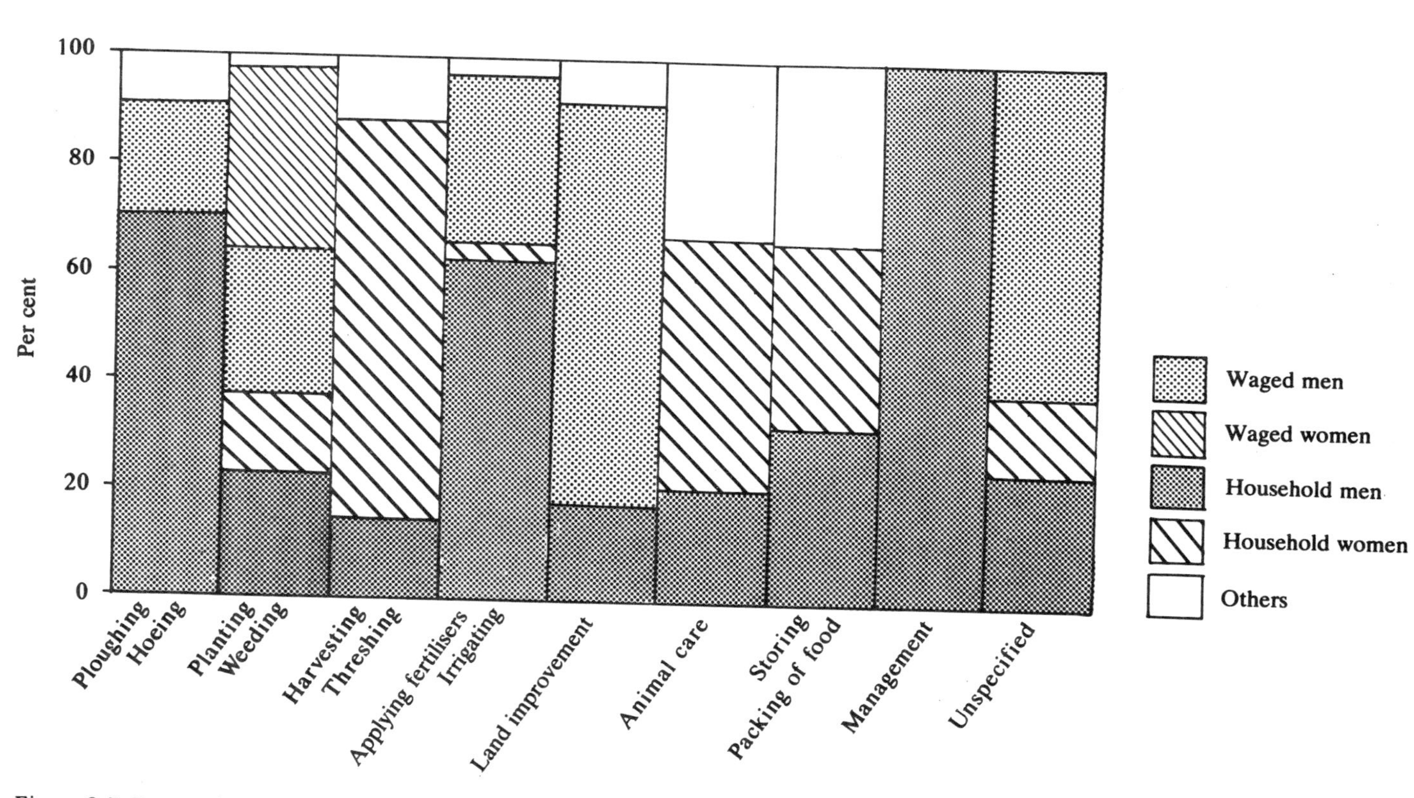

Figure 9.5 Proportion of total agricultural labour done by adult men and women – household and wage, slack period: April–May

significance in households with larger landholdings. There was no incidence of women ploughing in households owning over 10 acres of land. The time spent on agricultural work is much higher for women in poorer peasant households. Harvesting–threshing and sowing–planting–weeding–pruning follow the patterns of seasonality and decreasing contribution of female householders with increasing landholding size. Capital-intensive operations such as application of fertilisers, pesticides or irrigation water however, peak at the second- and third-lowest landholding sizes. These activities are of greater significance in larger landholdings with greater capital availability. In the largest landholdings, the contribution of women once again decreases as women gradually withdraw from farm labour. Heavy agricultural work is of increasing significance in the work schedule of women from smaller peasant households.

Inter-seasonal variability of total female labour use is higher for small peasant households than for larger landowners. The data on time spent per woman in agricultural activity, standardised per hectare, has a much greater range in the smallest landholding category. The small land-owning women contribute peak labour inputs on their own farm in March–April, which is a slack period for females in big landholding households. The women from small peasant households have a second peak labour demand time in the planting–sowing season in July–August. Their contribution to own-farm work in the harvesting–threshing season, the period of peak labour demands in middle-peasant households, is remarkably low, suggesting that the women seek wage employment in these activities in larger landholding households. The peak contribution of females in the largest landholding households is in the post-harvest season.

CONCLUSION

The analysis reveals the significance of female agricultural labour in the village of Karoli, India. Women were found to be working longer hours than men and more women worked on their own farms than did men. The gender division of familial labour in agriculture is thus biased towards the female adult. Invisible female agricultural labour in India needs to be disaggregated from 'family help' and accepted as a partner in agricultural production. This requires a realisation of the bias in data-collection techniques in large government organisations and a move towards correcting these biases. This must necessarily be preceded by more empirical studies.

Once women's significance in their contribution to agricultural production is recognised, variants in the labour pattern caused by seasonality need to be considered. Since the range of activities that women perform is limited, the problem of time imbalance arises between periods of peak female labour demand and the slack months. Only animal husbandry, which shows no marked seasonality in labour requirements serves to even out the work loads of women and provides alternative sources of income for female adult householders. Where the setting up of dairy co-operatives has led to the replacement of female labour with that of men, it not only causes female unemployment but increases the effects of seasonality of women's agricultural labour, which may be masked as the underemployment of women. State-sponsored employment-guarantee schemes need to take this into account.

Finally, given the heavy time demands of agricultural labour on women, there is a need to re-evaluate the social categorisation of males as 'providers' and females as 'housekeepers' and 'childrearers' and question men's low contribution to domestic labour in India.

10

ASSESSING RURAL DEVELOPMENT PROGRAMMES IN INDIA FROM A GENDER PERSPECTIVE

Rameswari Varma

INTRODUCTION

Feminist movements and the growth of 'conscientisation' of gender inequities and inequalities have introduced gender concerns into the development argument. The questions that are relevant in this context are: How have women been integrated into the development process? Has this process increased their status or increased their struggle? Has development had differential effects on women of different classes and castes? The need to recognise the implications of programmes and policies for the development of women and to integrate gender concerns in Indian development have been highlighted by several authors (Krishnaraj 1988; Sen and Grown 1985).

Among developing nations India has the distinction of a sustained development effort through systematic planning. But in spite of some forty years of planned development, women's status in India is far from satisfactory. On the whole, the development process has by-passed the majority of women, particularly in the rural areas. Over the past four decades women have been increasingly marginalised in Indian society. This is borne out by the declining sex ratio and its disturbing association with female mortality rates and female foeticide. Female labour-market participation rates have also decreased. Literacy rates among women have shown little improvement and the school drop-out rate is very high for female children. The health-care and nutritional status of Indian women leaves much to be desired (Neeradesai and Krishnaraj 1987). A large part of women's work in urban areas and even more in rural areas is 'invisible' and poorer women have to bear the multiple burden of work inside and outside the home and that of seeking out survival strategies (Jain and Banerjee 1986).

This chapter examines poverty-alleviation programmes from the woman's angle. Following a brief outline of the Indian government's policy towards development for women, it looks at the integration and share of women in the various poverty-alleviation programmes and assesses the impact of one of these programmes, the integrated rural development programme (IRDP), on women's status. The study is based on field research in Mysore taluk in the Mysore district of Karnataka State.[1]

INDIAN POLICY TOWARDS WOMEN'S DEVELOPMENT

'The evolution of policies towards integration of women in development has been slow despite the need for it' (Parthasarathy 1988). The first programme for rural development in India after the inception of planning was the community-development and rural-extension programme. Women were integrated into this programme almost as an afterthought. The approach toward's women's development was of a home–extension–welfare nature. In this programme women were taught some practical skills aimed at making them better housewives and using their time more fruitfully. Mahila Mandals in the rural areas were visualised as the catalysts for such development. But this programme did not meet with much success because it was a middle-class model for the promotion of women's welfare and the package offered was irrelevant to the needs of the rural masses (Mehra and Saradamoni 1983, cited in Parthasarathy 1988).

After the community-development programmes receded into the background, the focus of policy regarding women throughout the period of the Second to Fifth Plans (1955–80) was on welfare measures. In health and family planning, concerns about women have found explicit expression. However, it is only in the Sixth (1980–5) Plan document, which has a chapter on 'Women and development', that we see an awareness of the need for integrating women in development (Parthasarathy 1988).

The Seventh Plan (1985–90) displayed a greater concern and understanding of women's issues and proposed a multi-pronged approach to women's integration in development. In rural development programmes, and particularly in the poverty-alleviation programmes, the plan proposed that women should have at least a 30 per cent share. The plan exhibited concerns about women also in areas such as employment, health and nutrition (Government of India, Seventh Plan document 1985). As a prelude to the nature of integration of women in the Eighth Plan, the government of India has brought out a Draft National Perspective Plan for women which promises several measures for women's development.

A review of the policy towards 'Women and development' thus highlights two aspects: firstly that women and development was looked upon basically in welfare terms until the Sixth Plan period and secondly that it is only with the Seventh Plan that serious note has been taken of women as participants in development. But even in the Seventh Plan we do not find any radical measures directed at changing women's socio-economic status.

WOMEN AND POVERTY ALLEVIATION IN MYSORE TALUK[2]

There have been a number of studies on Indian poverty-alleviation programmes but not even a handful have attempted to evaluate the share of women in these programmes. One excuse for this failure is the shocking state of affairs regarding the availability of sex-segregated data for the poverty-alleviation programmes. It is only recently that data has been kept on a gender basis. Though Mysore district has eleven taluks or blocks, only Mysore taluk has maintained sex-segregated data. For other taluks, only since 1987 have women beneficiaries in some programmes been separately classified. To cull out sex-segregated data it was necessary to go through all the thousands of applications that were sanctioned from the files of the Zilla Parishad office and the Block Development Offices.

The object was to ascertain the share of women in the various anti-poverty programmes. This gives us an indication of women's participation and integration in these programmes. Many of the programmes involve women by providing assistance for the development of economic activities. We wanted to find out what was the nature of these activities so that we could get an indication of the ideology of the government towards women. The inferences presented here are based on such data as was available and also on discussions with the concerned Assistant Project Officers, the Block Development Officers and the Deputy Secretary of Zilla Parishad.

Regarding women's participation in IRDP Programmes (Table 10.1) we find the official booklet on DWCRA suggests that only 1 in 10 recipients of IRDP assistance are women. The Seventh Plan document states that on the whole women comprised 7 per cent of the beneficiaries covered under the IRDP during the Fourth Five-Year Plan (Government of India, Seventh Five-Year Plan, 1985). According to the District Rural Development Society's project officers in Mysore district, it appears that the official requirement that 30 per cent of beneficiaries should be women, has been taken seriously only since 1987. In the concurrent evaluation study of IRDP beneficiaries conducted by the Mysore Institute of Development Studies, out of 880 beneficiaries surveyed in Karnataka State only 160 were women.

Table 10.1 gives information on the various anti-poverty programmes

Table 10.1 Integration, share and nature of participation of women in the poverty alleviation programmes in Mysore taluk

Scheme	*Nature of scheme*	*Years data available*	*Total recipients*	*Women recipients No.*	*%*
IRDP	Income generation through assets	'81–'87	3,848	646	16.8
NREP RLEGP	Income generation through wage employment	Sex-segregated data not available			
Anthyodaya	Income generation through assets	'83–'87	657	84	12.8
TRYSEM	Training	'79–'86	2,671	1,185	44.3
Asthra Oven	Distribution of fuel efficient stoves	'84–'88	4,818	4,181	100.0
Biogas	Energy saving	'82–'87	2,028	140	6.9
Public Housing	Housing	'78–'82 and '83–'87	2,045	347	16.9
DWCRA	Income generation through economic activity	1988	–	–	–
Total (excluding Asthra Oven and DWCRA which are women specific programmes			11,249	2,402	21.3

Source: Compiled by the author

in Mysore taluk, the nature of the programmes and women's participation in each of them. Two inferences can be drawn: firstly, that women's participation has been minimal; and, secondly, that the nature of the economic activities chosen or allocated for women are limited to dairying, sheep and goat rearing, mat weaving, basket making and similar activities considered conventional women's activities. In the NREP and RLEGP[3] employment-generation programmes, women do not undertake heavy manual tasks but jobs such as carrying mud and breaking stones.

EVALUATION OF THE IRDP FROM A WOMAN'S PERSPECTIVE

The Integrated Rural Development Programme (IRDP), inaugurated in 1978–9, is one of the most prestigious rural-development programmes in India. It is the centrepiece of the anti-poverty programme in the Sixth and Seventh Plans. The target group consists of the poorest of the poor in the rural areas, that is, small and marginal farmers, agricultural and non-agricultural labourers, rural artisans and craftworkers, and people from the Scheduled Castes and Scheduled Tribes. The IDRP treats the whole family as a unit. The beneficiaries are families whose income is below Rs6,400, according to the Seventh Plan guidelines but the cut-off point for assistance is Rs4,800. The programme envisages providing needs-based assistance to the identified families in the form of assets by loans and subsidies in each block. The beneficiaries are selected according to the *anthyodaya* approach, that is by choosing the poorest of the poor for the first assistance. In addition, at least 30 per cent of the benefit has to go to the Scheduled Castes and Tribes. It is now proposed that 30 per cent of the beneficiaries must be women.

Under IDRP the block is taken up as a unit and the entire block (taluk) is covered in a phased manner. Maximum emphasis is given to the individual beneficiaries in order to make the individual schemes a success. Persons connected by blood or marriage or living together are considered to be a household for ascertaining incomes of selected target groups and a detailed household survey for each block is conducted. The IRDP is financed partly by credit from financial institutions and partly by subsidy from government. The credit requirements for financial institutions is about 2 to 3 times the amount of financial subsidy.

The total expenditure on implementation of the programmes during the Sixth Plan period amounted to Rs17,668 million as against the target amount of Rs15,000 million. As regards achievement, the target of 15 million beneficiaries was over-reached, with 16.6 million beneficiaries by 1985. The Seventh Plan proposes to cover about 20 million households under different schemes of IRDP. An outlay of Rs13,288 million has been provided for IRDP and allied programmes in the Seventh Plan in the Central Sector to be matched by an equal amount from the state.

Several evaluations of IRDP have been carried out (Rath 1985; Hirway 1985; Rao and Erappa 1987; Dantwala 1985; Dantwala *et al.* 1986) but most have been concerned with the nature of the programme, its philosophy, and its implementation, equity and efficiency. The critique of IRDP can be broadly divided into: firstly, implementation problems of corruption, malpractice and inefficiency; secondly, a questioning of the economic rational and viability of the programme because of the socioeconomic system within which it operates; and thirdly, the view that wage employment programmes are more suited to poverty amelioration than the provision of assets because the latter requires entrepreneurial skills, backward and forward linkages and regular payments of loans. The latter is one of the most important weaknesses of the programme. The reasons suggested for the high level of overdue payments at the banks involve the unsuitability of the schemes for the target groups, misuse of the asset acquired with loan, lack of linkages and insufficient entrepreneurial skills.

THE RESEARCH PROJECT

In contrast to previous evaluations, this chapter sets out to discover the share and nature of the involvement of women in IRDP, the performance of women in the programme and the extent to which participation has empowered women and influenced their perception of women's family role and image in society. In order to analyse women's position in IRDP a range of information is required, including the type and effectiveness of the assets acquired under the programme and the regularity of repayment. We also looked at whether women sought assistance from the programme on their own initiative, if they obtain assets themselves and whether they faced any harassment in obtaining the loan. The persons marketing the product funded by the loan and controlling the increased family income were also considered. The impact of programme participation on the women's work burden and health and on the workload of children and their ability to attend school regularly were also assessed.

The information was gathered in a survey of Mysore taluk through interview schedules and in-depth discussions. The choice of women interviewed was dictated by whether there had been at least 2 years between receipt of assistance through IRDP and the survey in order to allow adequate time for income to be generated. IRDP in Mysore taluk began in 1981–2. We were unable to get the particulars of women beneficiaries for the years before 1985–6. Therefore we selected women participants who were assisted during 1985–6 and 1986–7 and living in villages with at least two other beneficiaries. Some 60 per cent of programme participants were interviewed, 12 out of 18 who received assistance in 1985–6 and 38 out of 64 who received assistance in 1986–7 – a total of 50 women.

The women who were selected for this study belonged to the villages situated on the periphery of Mysore City. Most of these villages are directly connected to the city by bus transport. Such proximity to the city introduces elements of urbanisation into rural lifestyles and also increases employment opportunities.

The information collected was analysed for female-headed and male-headed households separately. In this study we have defined female-headed households as those households in which the main earner is the woman who receives assistance from the programme.

CHARACTERISTICS OF WOMEN BENEFICIARIES

The 50 interviewees were spread over 16 villages and included 18 female-headed households (36 per cent) and 32 male-headed households

(64 per cent). The average family had 6.3 members. Among women family members, 47 per cent were illiterate but 45 per cent had studied up to secondary level and 1.3 per cent up to college level. The literacy status of the beneficiary families was above the national average for rural areas probably because of the proximity of Mysore City. The age of all beneficiaries was between 30 and 60 years with most beneficiaries in their forties and fifties.

Of the sample households, only 22 per cent owned land. All the farmland was dry land and the largest holding was less than 1.2 hectares. Therefore, all the landholders belonged to the marginal and small-farmer categories.[4] Of the landowning families five were female-headed but the ownership of the land did not rest with them.[5]

Of the labour force of 221 in the sample, only 163 (95 men and 68 women) were engaged in income-earning activities outside the household. Amongst the men, 65 (68.4 per cent) were engaged in agriculture, 15 (15.7 per cent) in construction, and 4 in carpentry while 6 were in private enterprise, 4 were self-employed and one was in a government job. The female employment structure was much less diversified with 87 per cent engaged in agriculture and 13 per cent in domestic service. Identification of beneficiaries was done by village leaders (56 per cent), political parties (28 per cent) and the bank (16 per cent).

With regard to household income, half of the families had annual incomes below the poverty line of Rs6,400 per family. However, for purposes of IRDP assistance the cut-off line was set at an annual income of Rs4,800 and 38 per cent of the families were at this level. The employment pattern of the selected families in relation to income reveals that in each family earning above the cut-off point, several members of the family were earning a wage income. Since the city is nearby, a number of men have second jobs and a few also have a steady income working in private-sector establishments.

Though IRDP has several schemes to assist in income generation, only two have been extended to women in the sample: animal husbandry schemes (92 per cent); and assistance for petty commodity activities (8 per cent). As mentioned in the earlier section, the schemes provided under all programmes for women beneficiaries are limited to activities considered as belonging to the female domain. Most of the beneficiaries (82 per cent) obtained loans from commercial banks while the remainder have used regional rural banks. This is in line with the pattern for Karnataka.

Table 10.2 depicts the status of the income-earning asset acquired with the loan. For women beneficiaries 76 per cent of assets were intact or partially intact while the national average is 71 per cent (Kurien 1987). If we look at female headed households only, we find that 78 per cent of the assets given to women are intact. Clearly, the maintenance of assets by women beneficiaries is above average.

Table 10.2 Status and maintenance of loan asset

	Intact	*Partially intact*	*Not intact*	*Total*
Female-headed households	12 (67%)	2 (11%)	4 (22%)	18
Male-headed households	6 (19%)	18 (56%)	8 (25%)	32
Total	18 (36%)	20 (40%)	12 (24%)	50

Source: Author's fieldwork

IDRP loans have increased the incomes of the beneficiaries to varying degrees (Table 10.3). Female-headed households have the highest proportion of increases in excess of Rs2,000. Twenty-one recipients of whom 17 are men have shown no improvement in income generation because the initial asset is no longer intact. However, one of the problems that the IDRP has created is failure to meet loan payments due primarily to inadequate income and high household expenditure. Of the total number of programme participants, 22 per cent had repaid more than half of the loan and 38 per cent had repaid more than one-quarter of the amount borrowed. Although the loan repayment period was 4 years, nearly 60 per cent had paid more than a quarter of the total amount borrowed shortly after the receipt of the loan proceeds. Female-headed households tend to repay loans more promptly and more quickly; 44 per cent of the female-headed households had repaid 50 to 75 per cent of the loan and only 28 per cent had repaid less than one-quarter of the loan at the time of interview.

Seventy-two per cent of female-headed households sought development assistance on their own initiative compared to 59 per cent of the male beneficiaries. More than 61 per cent of the women household heads have themselves chosen the scheme of assistance. Half the female household heads but only a quarter of the male household heads went alone to the bank to receive their loans.

Management of the asset can be judged from Table 10.4 which shows that 39 per cent of women beneficiaries make repayments from income

Table 10.3 Increase in income as a result of loan asset (rupees)

Head of household	*0*	*501–1,000*	*1,001–2,000*	*2,000+*	*Total*
Female	4 (22%)	5 (29%)	6 (33%)	3 (17%)	18
Male	17 (53%)	5 (16%)	4 (12%)	6 (9%)	32

Source: Author's fieldwork

Table 10.4 Source of loan repayment

Head of household	*Family earnings*	*Income from loan asset*	*Both*	*Total*
Female	6 (33%)	7 (39%)	5 (29%)	18
Male	12 (38%)	13 (40%)	7 (21%)	32
Total	18 (33%)	20 (40%)	12 (24%)	50

Source: Author's fieldwork

from the asset and go to the bank alone to make these payments. The reasons for not going alone vary: for nearly 27 per cent it is because they feel they do not know how to handle business transactions, whereas others do not go because family members do not allow it or because they suffer from ill health. Women overwhelmingly (77 per cent) feel that the IRDP assistance has been useful to them. This compares with only 44 per cent of men. When women felt that the IDRP loan was not a useful source for income generation it was usually as a result of loss of livestock since most women extended loans to expand animal husbandry activities.

Respondents were asked about their perception of whether giving loans to women is a better practice than giving them to men. It was felt that, in general, women were better managers and did not spend money unnecessarily, but others felt that women lacked efficiency and time to manage loans. Only 22 per cent of the beneficiaries felt that the acquisition of the asset had meant more work but they did say that they did not mind as it brought in more income.

In more than half the cases which have reported an increase in income from the assets acquired under the IRDP, the control of the income and decision making rests with women: 78 per cent of women who are the head of their household reported having control over their income and decision making. Most women interviewed did not agree that there is gender discrimination within the family. However, there does appear to be some kind of sexual division of labour in work connected with the

Table 10.5 Division of labour connected with the loan asset (number of households)

Activity	*Female*	*Male*
Grazing animals	1	24
Feeding animals	8	7
Washing animals	12	3
Milking animals	18	4
Cleaning sheds	12	5
Selling milk	17	2

Source: Author's fieldwork

loan asset. Grazing the animals is mainly a man's work, whereas milking and selling milk is a woman's. Cleaning animal sheds and feeding is shared, but much of the burden is on women (Table 10.5). Milk is not sold to the Village Societies because the women thought that the Society price was not profitable for them. The inputs for the animals and other purchases were made by men. Among the petty traders, men made the purchases from Mysore City, but both men and women took part in selling in the shop in the village.

CONCLUSION

Our attempt to enquire into IRDP from a gender perspective has thrown up several pointers which can be useful in exploring issues relating to women and development. Development programmes have neglected women. About 48 per cent of India's population is female, but in IRDP and other such programmes, women have received less than 17 per cent of funds. Women should get an equal share in these programmes and the government has stated that women's share should be increased.

Most activities for women funded by these programmes are in traditional 'women's activities' that fit the conventional image of women. There has been very little effort to diversify activities. During discussions with Programme Officers, we found that though on paper the scheme of assistance is chosen by the beneficiary, in reality it is not so. Very often the woman is pushed to accept a scheme that is offered. Even in the schemes offered under DWCRA which is a group-oriented programme, meant only for women, this is the case. The involvement of women only in such activities that are in a traditional sense woman specific raises a few questions. Is this an indication of the strength of patriarchal ideology? Or is it a comment on the national 'development ideology' which has neglected the importance of rural diversification? Or is it a compounding of both factors? To what extent can mere increase in the quantity of women-specific activities help without planning for backward and forward linkages?

Our enquiry has revealed that women are efficient managers of IRDP assets. Their performance in maintaining assets, repayment of loans and in number of overdue payments is better than the national average. In all these aspects female-headed households are more efficient. Thus, there is a case for giving a greater share of funds to women beneficiaries in IRDP programmes.[6] This is further supported by the fact that the incidence of poverty, which these programmes are trying to address, is more prevalent among female-headed households (Parthasarathy 1988). That women beneficiaries have managed the assets efficiently, that they have been able to generate more income and that the entire increase in income generated by the women goes to the maintenance of the family strengthens the argument further.[7]

We began with the objective of trying to ascertain the effect of IRDP on economic and social status. The aim was to find out to what extent participation in IRDP as a beneficiary has 'empowered' women to function with autonomy. The concept of empowerment includes economic, social and cultural aspects. Under Indian conditions where there is a strong patriarchal ideology, where the roles of women are largely dictated by this ideology, and where women are not aware of the need to break this stranglehold, it is an uphill task for a researcher to find out whether any particular programme has 'empowered' women and/or improved her socioeconomic status. As an attempt towards this understanding, we asked questions related to the women's control over income, choice of scheme and general independence of decision making and spatial freedom. The answers to these questions revealed that although the assistance from the IRDP has not done great things by way of 'empowerment', there certainly has been some positive improvement in women's economic status and to some extent in her control over decision making. Women agree that IRDP assistance means more work but they do not seem to begrudge it and are positive in their attitudes towards IRDP. Since the type of assistance was mainly concentrated in animal husbandry, it did not have any harmful effects on women's health and may have helped family nutrition.

Though women have taken upon themselves the burden of the loan and have displayed a capacity for efficient management, the social perceptions of women do not seem to have been affected. Our discussions touched upon several aspects of the role and image perceptions of female beneficiaries. In this regard, the hold of patriarchal ideology seems very strong. Any programme to raise women's status should have a built-in component of fighting patriarchy and the value system that it produces. IRDP does not have any such component. We may conclude that assistance to women beneficiaries has improved their economic status but not empowered them sufficiently to break the shackles of patriarchal ideology.

It is true that improvement in economic status may not be a sufficient condition to change social, cultural and political status. Yet, it is our humble submission that it is one of the necessary conditions for raising the general status of women and all efforts towards it must be continued and reinforced.

Our study is essentially of an exploratory nature but our findings are certainly important pointers. It is necessary to investigate these issues on a larger scale and over a longer period to arrive at universally valid conclusions.

11

ACCESS OF FEMALE PLANTATION WORKERS IN SRI LANKA TO BASIC-NEEDS PROVISION

Vidyamali Samarasinghe

INTRODUCTION

With an annual per capita income of $380.00 Sri Lanka is listed among the 36 low-income countries of the world (World Bank 1988). However, for a poor country, Sri Lanka's record in providing a relatively high physical quality of life is known to be quite remarkable (Sen 1980). In terms of the Physical Quality of Life Index (PQLI)[1] as developed by Morris (1979), Sri Lanka is well above the level achieved by other low-income countries (Samarasinghe 1985). This is the result of egalitarian social welfare programmes, implemented by the state in Sri Lanka since the 1930s, that have attempted to provide its population with such basic needs as adequate food, access to education, health care and sanitary living conditions.

Aggregate statistics reveal that the gender gap in the physical well-being of the population is narrowing. Literacy rate for females was 82 per cent in 1981 compared to 90 per cent for males. The rate of increase in literacy of females has been higher than for males during the period 1946–81 (Samarasinghe 1986). Maternal mortality rate has declined from 15.0 per 1,000 in 1946 to 0.6 in 1987 (Department of Census and Statistics 1988). The findings of a fertility survey suggest that the country has entered the third phase of demographic transition identified with declining fertility rates (Department of Census and Statistics 1978). Despite the increase in the young reproductive age group (15–29 years) during the 1960s the fertility rate fell from 5.04 in 1963 to 4.22 in 1977, and now stands at 3.2 (World Bank 1988). The average age of marriage for females has been increasing, and by 1981 it was 24.4 years. Life expectancy for women in Sri Lanka was lower than that of males in 1963 but by 1981 had reached 71.7 years against 69.2 years for males.

However, if status is measured in terms of access to income-generating resources, the position of Sri Lankan women is low. Labour-force participation accounts for only 24 per cent of the total labour force. The unemployment rate is higher for females than for males and is three times higher for mid-level-educated females than for the corresponding category of males. As in many agricultural societies, women in the subsistence agriculture sector, though contributing much to food production, lie outside the official statistics (Samarasinghe 1986; Jayawardena and Jayaweera 1985).

Hidden within these national aggregates are deviant groups which display 'off-curve' characteristics. One particularly distinct group is the Indian Tamil plantation workers[2] of Sri Lanka who were originally brought to the island during the nineteenth century by the British colonial masters to work as labourers in the newly opened plantations. Compared to the rest of the female population in Sri Lanka, this group seems to have a lower physical quality of life, despite having better access to wage employment opportunities in the sector of the economy that has become the country's main foreign exchange earner. It is argued in this chapter that this group has been by-passed by the state-sponsored welfare schemes for reasons which are both historical and political. Furthermore, it is argued that although employment is high in the plantation sector, since the overwhelming majority of the women are labourers who receive relatively low wages, their standard of living is also relatively low. It is also the contention that the patriarchal cultural norms that pervade that particular society have led to an implicit subordination of women which has given them reduced access to basic needs.

THE INDIAN TAMIL POPULATION GROUP

The Indian Tamil population group accounts for 5.6 per cent of the total population of the country. They are mainly Hindus and speak the Tamil language. Seventy-four per cent of the Indian Tamil population live in plantations. More than 70 per cent of the tea plantation areas are in the central highlands of Sri Lanka, with a heavy concentration in the district of Nuwara Eliya, where a large number of tea plantations are located (Table 11.1, Figure 11.1).

The female Indian Tamil population has the highest rate of wage employment among females in Sri Lanka. The female labour-force participation rate in this sector which is dominated by plantation employment stood at 54.3 per cent in 1981. The rural non-estate female labour-market rate was 17.2 per cent for the same year. However, the women plantation workers have a far lower quality of life than the national figures for Sri Lanka would suggest. For example, maternal mortality in the main tea plantation areas where an Indian Tamil population predominate, is significantly higher than the Sri Lankan

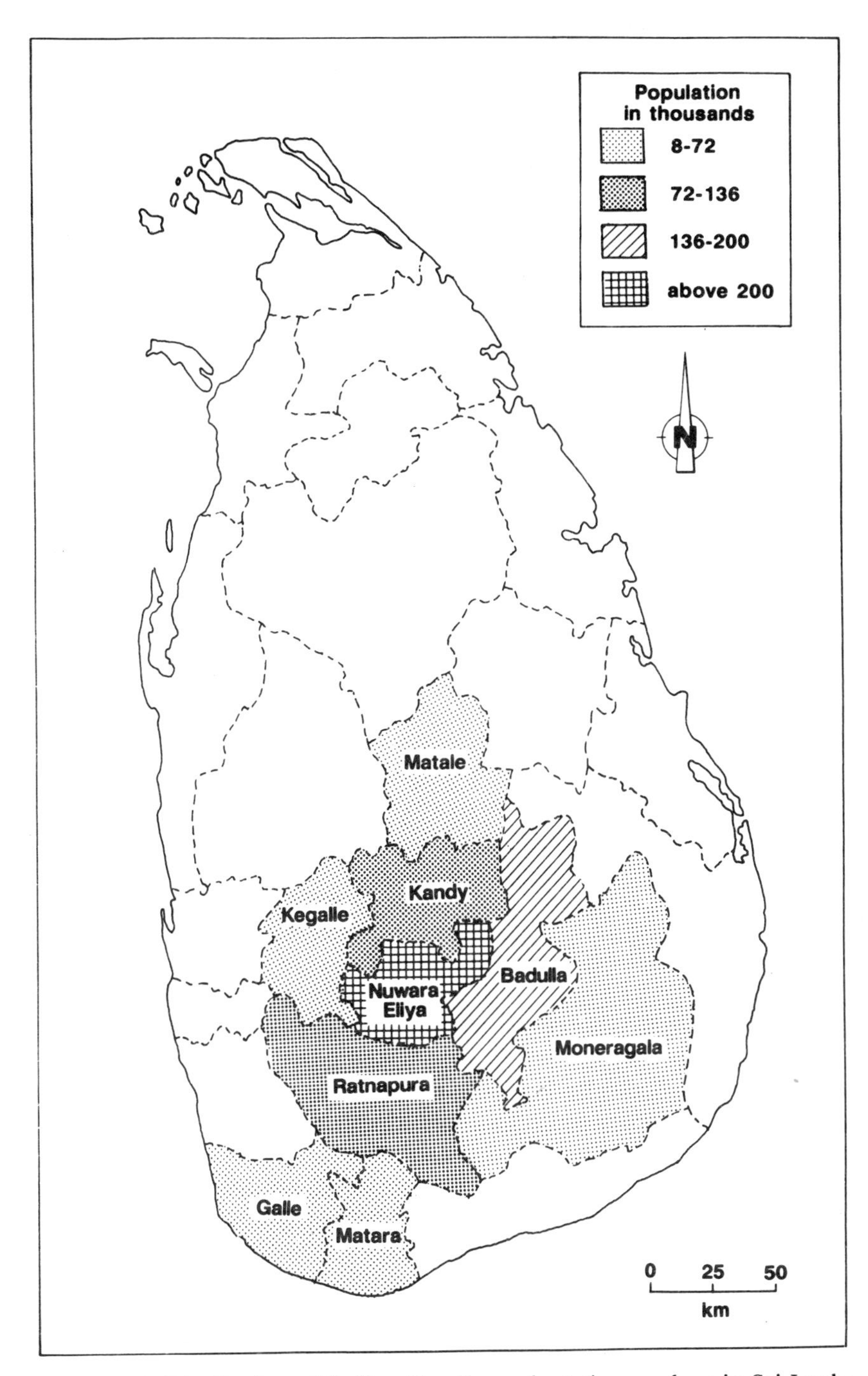

Figure 11.1 Distribution of Indian Tamil tea plantation workers in Sri Lanka

Table 11.1 Ethnic composition of the plantation population

District	Popn[a]	main crop	% Tamil I	% Tamil S	% Sinhalese	% other
Nuwara Eliya	312,657	Tea	74.0	18.0	6.3	1.7
Badulla	161,444	Tea	77.2	13.3	7.6	1.9
Kandy	130,495	Tea	64.6	15.6	15.3	4.5
Matale	29,777	Tea/rubber	60.5	24.7	12.3	2.7
Kegalle	54,908	Rubber[b]/tea	69.7	13.7	16.3	0.1
Ratnapura	107,569	Rubber[b]/tea	78.1	8.7	11.7	1.4
Kurunegala	10,920	Coconut	37.6	14.5	47.0	0.9
Kalutara	44,518	Coconut/tea/rubber	48.3	14.4	36.6	0.7
Matale	20,077	Coconut/tea/rubber	61.6	11.6	26.3	0.5
Monaragala	8,974	Tea	67.6	18.1	14.7	0.2

Source: Department of Census and Statistics, 1981
Notes: [a] total plantation population
[b] mainly rubber
I Indian Tamil
S Sri Lankan Tamil

Table 11.2 Maternal mortality rates of the main tea plantation districts where Indian Tamil population is concentrated (per 1,000 population, selected years)

District	% Tamil (Indian)	'72	'74	'76	'79	'80
Nuwara Eliya	52.3	2.3	2.4	3.1	1.4	1.0
Kandy	24.1	2.2	2.6	2.0	1.2	1.0
Badulla	17.8	2.0	1.4	1.4	0.9	1.0
Ratnapura	17.1	1.1	0.8	1.8	1.2	1.0
Kegalle	9.4	1.1	1.7	1.5	0.6	0.4
Sri Lanka	5.6	1.3	1.0	1.0	0.8	0.6

Source: Department of Census and Statistics, 1987

average in most areas (Table 11.2). Nuwara Eliya district has consistently recorded an above-average maternal mortality rate. It also has the highest percentage of Indian Tamil workers and depicts a similar pattern for infant mortality rates (Table 11.3).

The Indian Tamil plantation population records the lowest levels of literacy for the country (males and females). Only 66.9 per cent of this population group were literate as compared with 88.4 per cent for the Sinhalese and 86.9 per cent for the Sri Lanka Tamils. Among all males in Sri Lanka, the Indian Tamil group has the lowest literacy rate with 78.6 per cent while among females, the Indian Tamils record the lowest rate among all groups with 55.1 per cent. The widest gender gap in literacy among all ethnic groups in Sri Lanka is also seen within the Indian Tamil plantation group.

Table 11.3 Infant mortality rates in the main tea plantation districts (per 1,000 live births, selected years)

District	*'72*	*'74*	*'76*	*'79*	*'80*
Nuwara Eliya	85	119	100	79	74
Kandy	66	92	61	60	55
Badulla	59	73	51	57	47
Ratnapura	61	66	64	55	43
Kegalle	49	60	51	33	31
Sri Lanka	46	51	44	38	34

Source: Department of Census and Statistics, 1987

There has been a reduction of plantation Tamil population since the 1960s. This is due to three factors: the repatriation programme under the Srima/Shasti Pact of 1964, land reform induced reductions of plantation acreage since 1972 and a programme of land redistribution in the 1970s (see Peiris 1988; Rote 1986).

Historical background

The distinct characteristics of this group owe much to the genesis of the plantation structure itself, which rendered the immigrant labourers a 'separateness' from the rest of the population. Their ethnic identity is linked to a unique historical process of immigration, structured to feed the labour demands of large plantations. Immigrant labour was required to live within the perimeter of the plantations. Since the plantations were clustered together, the Indian Tamil population formed a concentrated residential pattern spatially segregated in the central highlands. They were separated by ethnicity, religion, language and economy from the native Sinhalese. They formed a separate 'enclave' associated with the export-oriented cash economy and had minimal interaction with the native Sinhalese, and their subsistence rice economy.

The method of recruitment of Indian immigrant labour during the British colonial period consisted of engaging labour gangs from poverty stricken, illiterate groups of people from Southern India. The Governors' dispatches during the colonial period consistently referred to them as 'coolies'[3] and their dwellings in the newly opened plantations as 'coolie lines'. It was not in the interest of the capitalist management structure to offer anything but the minimum standards of living for the immigrant labour. Since labour had to be residential, they were provided with barrack-type line rooms and minimum food rations (mainly rice) as part of their wages. Wages were kept to a minimum (Wesumperuma 1986) and health-care facilities were very poor. Education was not considered a necessity. At independence in 1948, nearly 100 years since their original recruitment, the Indian Tamils remained a polarised group, still identified with a low standard of living.

The political process in 1948 resulted in a disenfranchisement of this group, which number nearly one million. The system of electoral parliamentary politics in Sri Lanka, where members of parliament are accountable to their constituency, has been one of the main factors which led to a spatial diffusion of welfare facilities. Thus, the provision of free education from kindergarten through university and free health services benefited all citizens of the country who had voting rights. The Indian Tamil plantation workers, banned from voting due to the 1948 Citizenship Act, were effectively by-passed by the major welfare schemes. Instead, their welfare remained in the hands of plantation management. Since the management made no significant effort to match the welfare measures adopted by the state, the gap in physical well-being between the Indian Tamil plantation population and the rest of the Sri Lankan population widened. The Indian Tamil plantation workers remained 'stateless' until the 1964 Srima–Shastri agreement.[4] The last remaining numbers of the stateless Indian Tamil plantation population were offered Sri Lankan citizenship by legislation enacted in November 1988.

In 1975, with the nationalisation of estates under the land reform programme, the bigger plantations were brought under two semi-government organisations: the State Plantation Corporation (SPC) and the Janatha Estate Development Board (JEDB). More than 80 per cent of the Indian Tamil plantation workers are now employed by these two corporations.

Working conditions

Female tea plantation labourers are employed almost exclusively as 'tea pluckers'.[5] A typical worker's day begins before 4 a.m. in order to fetch water and prepare breakfast/lunch, clean the house and get children ready for creche and/or school. The morning meal is home-made bread (*roti*) with a watery curry and tea. These chores have to be completed before she reports for work at 7 a.m. The tea pluckers work in groups and keep filling the baskets with tea leaves until the tea break at 9:30/10:00 a.m. At this time, lactating mothers visit the creche to nurse babies. Resuming work after 30 minutes, each woman continues plucking until 12:30 or 1:00 p.m. when she takes the load to the weighing shed, visits the creche and goes home for the mid-day meal which she cooked the night before. She is back in the fields at 2:00 p.m. and plucks tea until 4:30 when she again takes the load to the weighing station and waits her turn. She visits the creche to collect the children and gets home about 5:30 in the evening. She begins the evening chores, cleaning the house, washing clothes, preparing the evening and following day's mid-day meal, feeding the children, cleaning them and getting them ready for bed. She is the last to go to bed around 10:00 or 10:30 p.m. She usually sleeps on a mat or a gunny bag spread on the floor. There is often only one string cot in the one-roomed line house which is used by the husband. She is free on Sundays and she is entitled to a few days' paid leave. However, during the main tea plucking season (May to July), she goes to work on Sundays and holidays in order to earn double the normal rate for 'over-kilo' plucking.[6] Her work involves climbing steep slopes, being exposed to rain, chilly winds and hot sun, and carrying a weight of up to 25 kilos in a basket strapped to her back.

Housing

Plantation labourers live in cramped accommodation known as 'line rooms', a long row of dwellings divided into several small units. Each unit would have one room (10 ft by 12 ft) with a small verandah. The plantation family size is estimated to be 4.8 persons per household. Since the early 1970s there has been a marginal improvement in housing conditions of the plantation workers – this appears to be related to a reduction in estate populations through repatriation rather than expansion of the housing stock (Peiris 1988). However, Table 11.4 clearly indicates that more than half of the plantation families have very limited living space.

Table 11.4 Plantation sector housing: residential units classified by number of living rooms

Residential units	*% of families per each unit* *1969–70*	*1981–2*
1 room	50.1	31.2
2 rooms	33.3	45.4
3 rooms	11.8	12.8
4 rooms	3.0	7.2
5 rooms	0.9	1.7

Source: Socioeconomic survey 1969–70, Department of Census and Statistics, Colombo, 1971 and Central Bank, 1984

The dwellings are often damp, cramped and the main living room tends to double as the kitchen. Cooking is undertaken indoors on open, firewood stoves without proper chimney ventilation. Consequently, smoke inhalation is a common health hazard. There are no ceilings (although in some households empty fertiliser bags are used to make a sagging ceiling) and the floor is made of cow dung and mud which the women have to resurface regularly. Rats and cockroaches are common. Furniture and implements are limited to string cots, a couple of chairs and a few brass pots. Drains adjacent to the houses are dirty while only 52 per cent of plantation households enjoy toilet facilities. The corresponding (1984) figures for the urban and rural sectors of Sri Lanka were 85 per cent and 68 per cent respectively (Central Bank, 1984).

In the JEDB Housing Needs Assessment Survey (JEDB 1986) it is envisaged that under the Medium Term Investment Programme (MTIP 1985–9) Rs200 million will be spent on improving housing needs. This includes constructing 1,250 new units and upgrading 6,000 units for the labourers.

Income

Until 1984 women plantation labourers received lower wages than their male counterparts despite working longer hours. This discrepancy was eradicated in 1984 with the equalisation of wages for male and female labour at Rs23.73 per day, per worker. Average monthly earnings for both males and females in the plantation sector have increased over the years (Peiris 1988). However, policy changes in the food subsidy scheme since 1978 have affected the plantation workers adversely. Food subsidy schemes implemented by successive Sri Lankan governments since the early 1970s granted a certain quantity of rice per person, either free of charge or at a subsidised price. This system was replaced in 1978 by a means-tested food and kerosene stamp scheme applicable to households with a total income level below Rs300 per month. Edirisinghe's (1986) study reveals that immediately prior to the food stamp scheme, the income support of rice subsidies was equivalent to 20.8 per cent, 21.4 per cent and 26.54 per cent of average per capita income in urban, rural and plantation sectors respectively. By 1981, these figures were respectively 13.3 per cent, 15.4 per cent and 10.2 per cent with the largest decline experienced in the plantation sector. Thus, although the plantation sector had received increased wages by 1984, this has to some extent been offset by declining income support, coupled with high inflation of food prices during the 1980s.

It has often been argued that women's income contributes significantly to the well-being of herself as well as her family (Kumar 1977; Guyer 1980). It is also assumed that women control the income they earn. However, the Tamil female plantation workers do not collect their own wages. They are usually collected by their husbands, fathers or any other adult male in the household – a practice which originated during the British colonial times. While it is true that Indian Tamil plantation women have relatively better access to employment opportunities, they do not seem to have effective control over their income. During the preliminary investigations of a study of maternal nutrition and health status of female plantation workers, young women reported that they do not have time to stand in line to collect their wages and that their husbands pass the wages on upon collection. However, plantation management personnel confirm Kurien's (1981) claim that frequent family conflicts arise because men tend to spend their wife's wages on gambling and alcohol. Inability to control one's own wages may effectively mean that the spending pattern of women wage earners on family welfare, as observed in certain other agricultural societies, may not be valid in the context of the Indian Tamil plantation workers in Sri Lanka.

Health, nutrition and maternity

The plantation management's obligation to provide for free medical care of their labourers was placed on a legislative footing in Section 27 of the Medical Ordinance no. 11 of 1865. This entitled plantation workers to lodging, medical care and food at the expense of the employer. However, while each plantation continues to provide a dispensary manned by a medical practitioner, not all plantations have qualified medical personnel, known as Estate Medical Assistants (EMA).[7] The dispensary stocks a limited quantity of medicine and seriously ill patients are transferred to state hospitals, admission to which has to be certified by the plantation superintendent who is expected to pay a nominal fee for each patient. In a medical emergency ambulance service is available in only a few plantations. Elsewhere, plantation workers can call for the estate lorry to transfer serious cases to hospital, although it can take up to 30 minutes to collect a patient due to difficult access to line rooms. Moreover, labourers dislike going to hospital because they feel isolated and lonely when away from familiar surroundings (Kampschoer *et al.* 1983).

Indian Tamil female plantation workers' tight work schedule and ignorance are major causes of poor health. Protracted exposure to inclement weather, poor diet and living in damp, ill-ventilated dwellings contribute to the incidence of respiratory illnesses. Furthermore, according to a Central Bank survey, diarrhoea and abdominal complaints were most pronounced in the estate sector as a result of polluted drinking water and lack of personal hygiene.

The author's personal experience, supported by the findings of another group (Kampschoer *et al.* 1983) showed that the women paid little heed to certain rudimentary factors of health care. In recent times, health educators through a series of health clinics have explained to the Indian Tamil female plantation workers the importance of drinking boiled cooled water. They did boil the water, but mixed it with unboiled water to cool it before drinking! The habit of using toilets is not common at all. Management constructed common toilets and the plantation workers claim that since no one was responsible for their cleanliness they became unusable (Kurien 1981). Another reason given was a lack of available water for the toilets. By 1987, water on tap was made available to most line dwellings under the management of the JEDB and the SPC. However, the drains around the houses were still used as latrines. The management has now initiated a programme of providing toilets for each family. While some of these are now being used, some others are used to store firewood. The basic problem is an inability for people to understand the health hazards of unsatisfactory hygiene.

Low incomes have often been cited as a cause of low nutrition among plantation workers (Rote 1986), particularly during the British colonial period (Wesumperuma 1986). However, incomes have increased since nationalisation in 1975 (Peiris 1988). Edirisinghe (1986) has used the concept of the 'ultra-poor' (Lipton 1983) to study the efficacy of the food stamp distribution system in Sri Lanka. The 'ultra-poor' households were defined as those failing to achieve 80 per cent of the recommended calorie allowance despite allocating over 80 per cent of their expenditure to food. Among the agricultural population of Sri Lanka, the plantation workers are not at the bottom of this hierarchy. Therefore, given their level of income, plantation workers could better afford the minimum calorific intake as compared with other agricultural workers. But what is noteworthy is that although the per capita calorie intake among plantation workers and their families was about 2,000, which is one of the highest among all groups in Sri Lanka, the rate of chronic undernutrition in the plantation sector was 60 per cent; in the urban sector it was 33.6 per cent and in the rural sector 33.2 per cent (Janatha Estates Development Board 1984). In seeking an explanation for this, certain factors need to be taken into account.

Food preparation in the plantation households is exclusively in the hands of women. However, their tight wage-work schedules allow little time for food purchase or preparation. Levels of household technology available to plantation women are very low and traditional food preparation techniques are time consuming. A typical meal (dinner) consists of rice and one curry. Being Hindus, some of the Indian Tamil population avoid eating beef, but this is not a strictly observed practice. The real problem is inaccessibility to a variety of food items, given the time available for cooking. The basic food items such as flour and rice are provided by the management and the cost is deducted from the wages. To purchase other items of food, women have to visit the grocery store in town. The small shop typical of a plantation does not stock a variety of food items and what are available are relatively more expensive than in a larger store in town. The physical location of plantations makes access to such shops difficult and women do not have the time to visit them anyway. Since fish or meat cannot be stored for lack of refrigeration, such items are included in the diet only very rarely.

Equally important is the ignorance of women with respect to the nutritional value of specific food items. Improvement in women's educational level has, in the recent past, been increasingly appreciated for its central importance as a determinant of family health and nutrition (Buvinic *et al.* 1988). Plantation women have very low literacy levels, and in the absence of a continuing programme on nutritional education, they cannot be expected to adhere to norms of nutrition in the selection and preparation of food.

Maternity benefits were also laid down by the Medical Ordinance roll of 1865, whereby labourers were entitled to a cash payment, quota of rice and the services of an attendant (Chatopadya 1979). It was also stipulated that there should be lying-in-rooms and trained midwives in plantations. In reality, while maternity leave and cash payments were granted for the permanent cadres of female labourers other facilities were at best only minimal. Even maternity payments were collected by the husbands and it is doubtful whether the money was used to purchase supplementary feeding or any other benefits for the lactating mother, or the new-born baby.

Maternal health benefits are currently an improvement on this situation. Indian Tamil women are now entitled to 12 weeks of maternity leave with pay, while most plantations have a maternity ward with four or five beds and a qualified midwife. Certain deductions are made from maternity benefits for expenses incurred during the stay in the maternity ward. From the third month of pregnancy the women are expected to report to the midwife. Visits are now more frequent since they are tied up with maternity benefits and a supplementary feeding programme. Supplementary feeding known as *triposha* which is a nutritious soya-based flour mix distributed free by CARE, is given to pre-school-age children, infants and lactating mothers.

Although women are expected to come to the maternity ward for deliveries, many babies continue to be born at home with the help of older relatives, especially for the first child, since it is a tradition that the first baby should be delivered by the young mothers' own mother at home. This is dangerous both to the new mother and the baby because primitive methods are used. Complicated cases are sent to the state hospitals; however, many wait until the last minute which results in unnecessary casualties. Tradition still plays a decisive role in pregnancy, child birth and child care, despite health education programmes and supplementary feeding schedules. Pregnant women in labour are given decoctions and alcohol, and it is common knowledge that up to four weeks after delivery, new mothers are given alcohol.

Researchers have noted that despite the relatively young age of marriage of Indian Tamil estate women, their fertility level is low. Langford (1982) notes that this may be due to a higher incidence of foetal loss, as a result of hard work, low nutrition and low health care. Ratnayake et al. (1984) attributed low fertility to longer periods of breast feeding. Caldwell et al. (1987) speculate that it may be also due to higher rates of abortion on the plantations. Modern methods of contraceptives, such as the Pill, IUD or the Depo provera injections are not used in this sector. Rather, tubectomy is the most popular form of contraception. Family planning is not geared towards child spacing but rather towards terminal methods after the birth of three or four children. It is evident that Indian Tamil plantation women desire to limit their families. They carry a heavy double burden. Younger women of child-bearing age are keen on obtaining a tubectomy because they are ignorant of other modern methods of contraception. They are also attracted by the cash incentives offered to those who will undergo a tubectomy.

Since the management of the bigger plantations has been taken over by the SPC and the JEDB, a plan for upgrading health care has been undertaken. A comprehensive health plan has been initiated under the Medium Term Investment Programme (MTIP) funded by the Asian Development Bank. A UNICEF-sponsored programme for health education, supplementary feeding, pre-natal, post-natal care and child care has been initiated. Although there has been a visible improvement in the infrastructure provided for health care in the plantation sector (Marga 1983) and a reduction of maternal and infant mortality rates, there are still certain serious limitations on the delivery system of health care as well as on the absorptive capacity of the plantation works to make use of them. Better health care given in the state sector is still physically inaccessible to many plantation workers. Health personnel in the plantation rely more on curative rather than on preventative methods of health care.

Education

Poor access to education for Indian Tamil plantation workers in general, and women in particular, owes much to the historical processes of labour recruitment, management perceptions, and the political process that rendered them 'stateless' until recently.

It was against the interests of the capitalist management structure during the late nineteenth and early twentieth centuries to offer educational facilities above the very rudimentary levels to the Indian Tamil plantation workers (Chatopadya 1979; Gnanamuttu 1977; Wesumperuma 1986). The 1905 Wace Commission Report on elementary education did not consider it desirable that there should be provision for general education in plantation schools. 'If we give the labourer the education which will fit him to be a clerk', the Commission observed, 'the results are naturally bad for the community' (quoted in

Chatopadya 1979). This sentiment has not changed much over the years. Despite the Education Ordinance no. 1 of 1920, where compulsory attendance at school was stipulated up to age 16, it was never implemented in the plantation sector. Although Tamil plantation schools were given certain facilities, schooling was limited to primary grades. By age 14, a student should be in grade 8 or 9. However, no estate schools had grades beyond the primary level, and 'school' consisted of one large hall, manned by one or two ill-trained teachers. Whoever wanted post-primary education had to trek long distances to reach schools in towns.

By 1955 only 67,000 children of the plantation sector attended schools. Among them only 25,000 were girls. There were a little over 800 plantation schools (Chatopadya 1979). By 1976 the number of estate schools had been reduced to 725 and there were 50,816 students (Bastian 1983). By 1980–1, while the age-specific school-avoidance rate for age group 5–9 years in Sri Lanka was 19.87 per cent it was 40.74 per cent for the plantation sector (Central Bank 1984). The index of education computed by the Central Bank was lowest for the plantation sector and when disaggregated by gender, the Indian Tamil female segment records the lowest score (Central Bank 1984). During the land reform of 1975 the estate schools were taken over by the state. However, a satisfactory plan to upgrade the schools and give easy access to plantation populations has yet to emerge.

Apart from poor facilities for education, other factors tended to discourage the participation of young girls in the education system. Female children were often kept at home to look after younger siblings while the parents were at work (Kurien 1981). Furthermore, young females were able to get ready employment by the age of 14 years as tea pluckers. Hence the levels of literacy remain low for Indian Tamil female plantation workers.

Education forms an important strategy for the upward social mobility of the people of Sri Lanka. Aggregate national statistics suggest that gender disparities are fast disappearing in relation to access to higher levels of education and that more and more women are seeking prestigious jobs (Fernando 1985; Jayaweera 1985). Education imported through a system of formal education could effect plantation women in two ways, by helping them break away from the inhibitory constraints imposed by traditions regarding health and nutrition, and also by encouraging employment aspirations beyond tea plucking. However, employment in other economic activities in these districts is very limited, and emigration of plantation workers out of this area is quite low. Being Tamil speaking, the outmigration pattern has been limited to certain pockets of Tamil-speaking northern agricultural areas. This phenomenon is no doubt influenced by a sense of insecurity that surrounds moving away from 'familiar areas', especially in the face of the prevalent ethnic strife in the country. Improved education is likely therefore to help swell the ranks of mid-level-educated unemployed among females in Sri Lanka.

CONCLUSION

The Indian Tamil plantation society is very patriarchal in nature with males dominating many aspects of decision making. Kurien's (1981) study shows that men had no sympathy with the extent of women's employment and domestic commitments. Men insisted that women went out to work; they scolded them if they were not at work on time and demanded that they worked extra hours, while avoiding household tasks themselves. The patriarchal nature of Hindu culture also has a decisive influence on the subordination of women in this community (Kurien 1981).

It is often argued that in traditional agricultural societies males are given priority treatment within the household especially with respect to food allocation, because it is believed that they need the energy to work hard in the fields. In the plantation sector, however, women work harder and there is no evidence to show that any priority is given to them in food allocation or in household decision making. The low level of female education undoubtedly contributes to the continued acceptance of patriarchal norms in the society.

The plantation labourers are represented by a very strong trade union movement. They have successfully negotiated a series of collective agreements on work schedules, wage hikes, work quotas etc. However, while more than 50 per cent of plantation labourers are women, all union leaders are males and there is little female representation. Better access to basic needs such as schooling, widespread health care and sanitary living conditions seem to receive low priority in union demands. The most successful union demands during recent times have been to ensure 26 days of work per labourer per month, wage increases from a daily rate of Rs9.91 for males and Rs6.73 for females in 1978 to Rs23.70 for both men and women in 1986. While enhanced incomes for women would theoretically increase their access to better living conditions, in reality they may not control that income at all due to the prevalent practice of husbands collecting the wages. There has been no demand by the union to get management to pay wages directly to each individual worker.

Furthermore, given more work and no notable decrease in her household responsibilities, she may well suffer from over-work. There are no facilities within the plantation sector to purchase cheap cooked food. Household technology remains primitive and time consuming.

Allocation of time among increasing wage and household work becomes a constant balancing act, and it is very likely that the woman will be forced to spend less time on obtaining and providing basic needs for herself and her family.

12

WOMEN AS AGENTS AND BENEFICIARIES OF RURAL HOUSING PROGRAMMES IN SRI LANKA

Yoga Rasanayagam

INTRODUCTION

A critical analysis of the role of women as agents and beneficiaries of state-sponsored rural housing programmes in Sri Lanka forms the major objective of this study. The analysis is restricted to the Aided Self Help programmes, the Model Village programmes and the Village Reawakening programmes under the Million Houses Programme, initiated by the present government. Sample households from each type have been chosen from the following districts: Kalutara, Colombo, Kurunegala, Kegalle and Moneragala (Figure 12.1). The size of the samples had to be restricted due to the prevailing troubled situation in the country.

RURAL HOUSING PROGRAMMES

Nearly 72 per cent of the population of Sri Lanka live in the rural areas. The majority of these people are engaged in the cultivation of paddy and other food crops on the lowlands, or in manual work, mainly in agriculture or fishing. A substantial proportion of the poorer people in the island belong to these groups and therefore many of the rural development programmes implemented to date have been designed to improve their income levels and elevate their living standards. Since the late 1970s housing has played a pivotal role in the development process.

The initial supply-oriented housing programmes were largely agricultural-cum-settlement based and were directed towards migration of landless peasants to new and undeveloped areas. In these programmes women-headed households were never considered. The objectives of the subsequent state-sponsored housing programmes of the 1970s, however, were much more broadly based. Housing was considered vital for

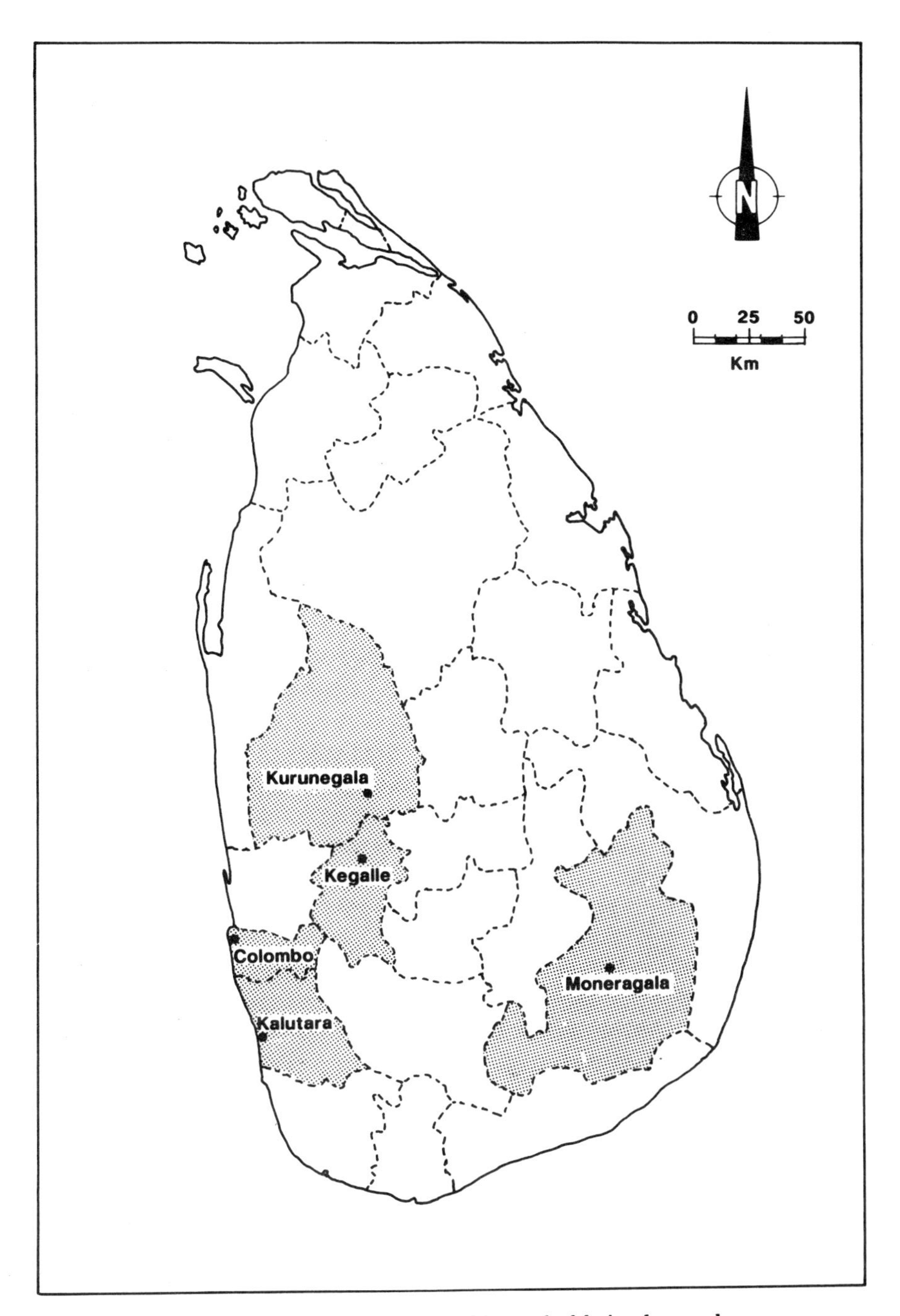

Figure 12.1 Districts of origin of households in the study

the development of the poor and homeless. Rural areas received priority in these plans. Initially, these programmes had both a direct state construction component and an Aided Self Help component.

Under the Aided Self Help programme introduced in 1972, the state met the cost of land development and certain services and building materials while the allottees provided the labour. In these instances family labour and assistance from neighbours included female labour, largely in the form of light work related to carrying bricks and cement and looking after the food and refreshments needs of other workers.

The Aided Self Help (ASH) approach enabled the government to effect considerable reduction in the cost of low-income housing units and spread the benefits of public-sector investment over a larger number of beneficiaries (Marga 1986) because an ASH unit cost only Rs6,000 as against the Rs30,000 per unit under the direct construction scheme. The ASH scheme also successfully tests the validity of self-help as a method of low-income housing. It encouraged community participation, wherein women as contributors or as agents supplied a portion of the unpaid labour component. Published data on the participation of women in building construction work as provided by the census reports (Government of India 1981) includes labour inputs by women in the construction of roads and canals as well. However, a clear assessment of unpaid female labour input in rural housing is not possible, and their contribution goes unrecognised, by default.

After the present government of Sri Lanka came to power in 1977, housing became one of the three most popular and prestigious 'lead' projects of the country. The prime minister started a scheme in 1978 of building 100,000 houses in four years. Of these, nearly 50,000 units were constructed on the basis of Aided Self Help in the rural areas. In this new concept of housing, the agricultural settler became the housebuilder–settler. The emphasis was on developing the villages in order to develop the country. There were several components to this rural housing programme namely, the ASH programmes, Model Villages scheme and the Electoral Housing scheme (see Rasanayagam 1988).

The Model Village scheme was the first stage of the village reawakening movement. In this programme, housing itself, initially built by the state and subsequently by the allottee with state assistance, is supported with activities which improve the social infrastructure of the village and at the same time promote income- and employment-generating activities. Although the model village programme is similar in many respects to the regular ASH programme, it does not select individual beneficiaries on income criteria. The entire village becomes the beneficiary when a village is selected for the programme. In this strategy, therefore, it is not the need that entitles a beneficiary family to a house but rather its membership in the community. Families which are not in need of state assistance also enjoy the benefits. This programme provides new houses with modern infrastructure and facilities as a replacement for existing housing units of poor quality.

The electoral housing programme, on the other hand, was initially implemented through a direct construction scheme. Consequently, it proved to be very expensive. The target was five houses in each electorate in 1978, increased to 50 by 1983. In 1984, a new strategy of development through housing was launched by the prime minister. This plan aimed at completing one million housing units by 1993. The philosophy of this new programme is to reach greater numbers of the poor and needy at less cost to the nation, but with more satisfaction to the beneficiaries. With state intervention reduced to the minimum, people's participation is encouraged to the maximum. Rural housing is an important sub-programme within this plan.

The Million Houses Programme (MHP) is an open-ended system in which the beneficiaries can decide and adopt standards incrementally in accordance with their needs. This is referred to as the Re-awakened Village system wherein entirely new houses need not be expected as in a model village system. Houses under the Re-awakened Village system of the Rural Housing Sub-Programme (RHSP) may be spread over a whole village or within a cluster of several villages.

WOMEN AS AGENTS OF RURAL HOUSING PROGRAMMES

Women's role in Sri Lanka as planners and decision makers in housing programmes remains neglected and unfulfilled. In the designing of houses for these programmes Sri Lankan women have not participated as members of planning boards or panels. As decision makers in the process of house construction, however, there is some evidence of there being a role for them in certain localities and in certain types of state housing programmes. Moreover, a comparatively large proportion of women contribute to the housing programme in the form of unpaid labour. Female members in almost all households selected in this study have contributed significantly to the labour input required for house construction. In the case of women heads of households, participation has been found to be disproportionately higher making it too exploitative of women. As seen in the statistics of sample households in the Self Help schemes, women heads of households lack the time and skills required, and therefore have to employ others to build the house for them and this has often been found to be very expensive.

WOMEN AS BENEFICIARIES OF RURAL HOUSING PROGRAMMES

Women as beneficiaries in these housing programmes are categorised, in this study, as follows:

1 women beneficiaries as household heads;
2 women beneficiaries as members of households with a male household head.

Table 12.1 indicates that female-headed households in the different districts of Sri Lanka vary from 11 per cent to more than 22 per cent of the total number of households in any district. Since 1981, however, when the last census was taken, the number of such households has increased several-fold due to the troubled situation in the country and also the prevalence of outmigration of male members. Table 12.2 indicates the number of women-headed households in each of the rural

Table 12.1 Number and population of female-headed households by district, 1981

District	*Total number households*	*Female-headed households number*	*%*
Sri Lanka	2,721,514	476,841	17.4
Colombo	277,721	51,721	18.4
Gampaha	255,747	43,954	17.1
Kalutara	155,013	29,789	19.2
Kandy	202,995	37,798	18.6
Matale	66,497	10,336	15.5
Nuwara Eliya	106,492	16,998	15.9
Galle	148,205	33,181	22.3
Matara	115,085	25,347	22.0
Hambantota	75,242	14,554	19.3
Jaffna	149,685	30,561	20.4
Mannar	20,057	2,179	10.8
Vavuniya	18,152	2,314	15.7
Mullaitivu	14,586	1,604	11.0
Batticaloa	62,843	12,088	19.2
Amparai	68,767	10,836	15.7
Trincomalee	44,661	5,667	12.6
Kurunegala	240,525	39,917	16.5
Puttalam	100,738	17,454	17.3
Anuradhapura	108,909	13,402	12.3
Polonnaruwa	45,452	5,826	12.8
Badulla	119,984	19,104	15.9
Moneragala	43,802	5,140	11.7
Ratnapura	150,249	22,089	14.7
Kegalla	130,184	22,449	17.2

Source: Based on 10% sample of Census of Population and Housing 1981.

Table 12.2 Women beneficiaries in rural housing programmes in Sri Lanka (by district)

District householders	*Total number of householders*	*Women number*	%
Colombo	2,008	310	15.4
Kalutara	2,783	358	12.9
Gampaha	2,789	508	18.2
Kandy	2,286	231	10.1
Matale	913	113	12.3
Nuwara Eliya	1,900	344	18.1
Galle	2,768	594	21.5
Matara	1,605	257	16.0
Hambantota	1,460	136	9.3
Jaffna	1,348	326	24.2
Mannar	316	39	12.3
Vavuniya	272	12	4.4
Mullaitivu	220	23	10.5
Batticaloa	1,275	236	18.5
Amparai	na	na	–
Trincomalee	632	44	6.9
Kurunegala	3,927	595	15.2
Puttalam	1,751	229	13.1
Anuradhapura	1,935	215	11.1
Polonnaruwa	859	119	13.9
Badulla	1,963	379	19.3
Moneragala	981	161	16.4
Ratnapura	1,721	187	10.9
Kegalla	1,359	229	16.9

Source: National Housing Development Authority, Sri Lanka (unpublished)

housing schemes in the different districts. Although such households are very limited in certain districts, it is heartening to note that the housing authorities have made a very salutary move by considering women heads of households among the needy. These women now own the small houses, although some of them find it difficult to pay back loan instalments, as these houses were built on the basis of self-help.

A closer analysis of six women-headed households in ASH schemes found that women-headed households faced immense problems in protecting building materials from thieves, in procuring the required building materials from available sources and in coping with the numerous burdens of looking after children and building a house. Since none of them had any knowledge of house design, the local mason (*bas*) played a dominant role except in the case of one woman who obtained the services of a qualified planner. The area of land allotted being small resulted in the space within the house being limited even though the planning was done by the allottee through the mason. In the five cases where the mason decided on the plan of the house, rear verandahs were not included in the building, thus making it very inconvenient for the women who prefer to spend most of their time doing household chores in conditions of privacy. In three of the households, toilets have not been included in the plan and with finances running low and the women unable to muster additional resources to build toilets they are forced to relieve themselves outside the premises at great inconvenience.[1] Drinking-water facilities are being provided by common wells which are not adequate for the scheme.

Under the Aided Self Help programme, the plan of the house to be built was provided by the Housing Authority along with the necessary building materials, leaving hardly any choice for the allottee in the design of the house. The male household head was the owner of the property. Families with many members had to manage within the small space provided for the whole house (plinth area approximately 56 m^2). All respondents have commented that the space of the entire house, let alone the rooms, was woefully inadequate. A male household head in such a house in the Moneragala district made some alterations in defiance of the wishes of his wife and daughters. He was able to add a rear verandah and a small extension on the side of the house. The expenses for this had to be borne by the allottee.

Nearly all women in these houses commented that the kitchen space was hardly adequate and due to the design of the house all the cooking smells, and particularly smoke, entered the house causing immense discomfort, particularly to the women who remained indoors for most of the day and night. The building materials that are given for the construction of the house have apparently been of very low quality; the roof of one house in the Moneragala district collapsed, barely missing the women who were indoors at the time. A new roof had to be built by the allottees incurring additional debt.

In terms of amenities, the case studies in the Moneragala district indicate that toilet facilities and drinking water were available to the allottees. But since only two tube wells serve 40 households in the scheme, women have immense problems in the collection of water. They have to endure long waiting times to obtain water, because some residents use the wells for other purposes such as bathing, even though they are specifically meant for drinking water. Though some women find the get-together at these tube wells welcome for gossip, the respondents in the study indicated that unnecessary quarrels begin at these wells sometimes leading to serious fights among the allottees. They, therefore, felt that more tube wells should be made available to avoid such undesirable situations.

Women in such a scheme in the Kalutara district seemed to suffer without proper wells, i.e. neither separate wells nor common wells. They had to share unprotected wells in the area. The social environment in this scheme too is found to be unpleasant and the women folk have to either keep indoors or go out of the area to get away. Inadequate space in the houses is a further constraint on development activities in the area.

Women in the model villages selected in the Kalutara, Moneragala, Kurunegala, Colombo and Kegalle districts have similar views on the design of the house (Figure 12.2). Here again, male heads of households are the owners and two of them built additional rear verandahs to provide extra space for the women in the house. Except for the model village selected in the Kurunegala district, (the first model village in Sri Lanka), all the others have been built by the allottees on a Self Help basis. Women in the Kalutara district expressed the opinion that the location of the house was so inconvenient that they found it difficult to use the living room during the rainy season because wind and rain beat into the house causing the additional problem of having to sweep the rain water out. The front door too had to be replaced at a cost that the family could not afford.

Respondents from the Moneragala district commented that because the land on which the housing scheme is built is low, the area around the house is flooded during the rainy season, resulting in water logging and leading to terrible inconvenience to the people in the house. The women residents apparently find their sufferings increased by the additional chore of keeping the house clean and dry with people walking in with wet feet. The children suffer from skin ailments during these periods and this again adds to the responsibilities of women. The consensus of opinion among most women in the households within the model village was that because the houses are built too close together the women could hardly have any privacy, let alone any space for income-generating activities.

All respondents in the ASH schemes and Model Village schemes have only been able to complete at least the minimum liveable house with financial assistance from friends and relatives because the loans given by the housing authorities were inadequate. Some families find it difficult to repay the monthly instalments of the loan. The women feel that they could contribute towards these instalments if they could find employment.

In terms of living space, the houses constructed under the Reawakened Village system appear to be satisfactory, particularly for the women. Some houses were built on land owned by women. Houses of all respondents under this system have been designed or planned by the householder or by the mason on the instruction of the householders. In one case, the wife of the male household head was the decision maker and adviser for the plan. This woman was enterprising enough to seek ideas from model villages nearby. She visited people living in a nearby

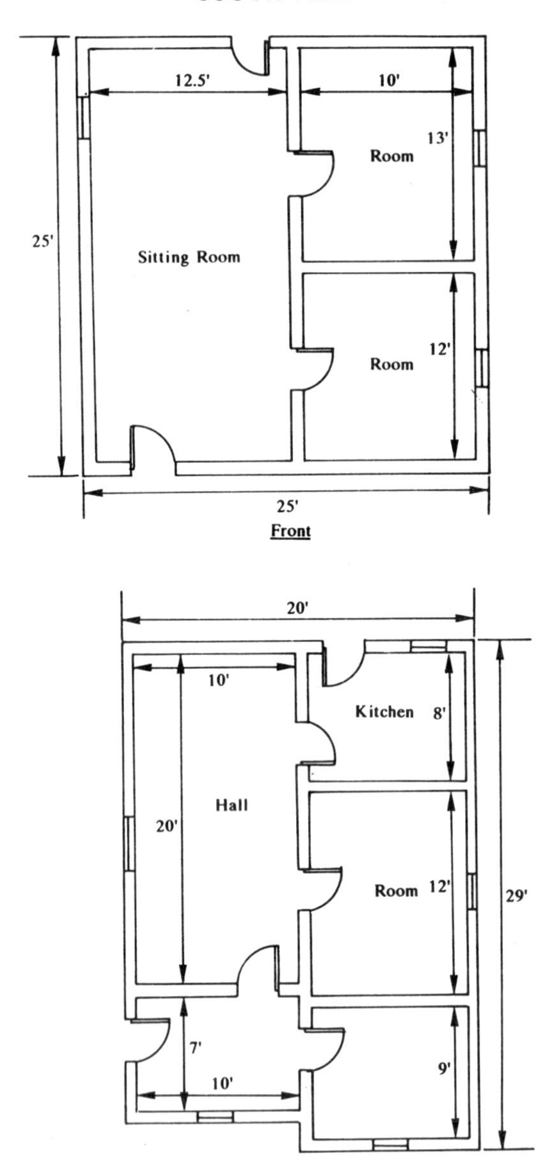

Figure 12.2 House plans 1 and 2, reawakened village

model village to study the design, space availability and construction requirements. As she found the space of the model village house limited, she instructed the mason who helped to draw the plan to increase the dimensions of the living room and the bedrooms to suit the family (Figure 12.3 and Figure 12.2 plan 1). The kitchen was built outside the house. In another case, however, the designing and planning was done

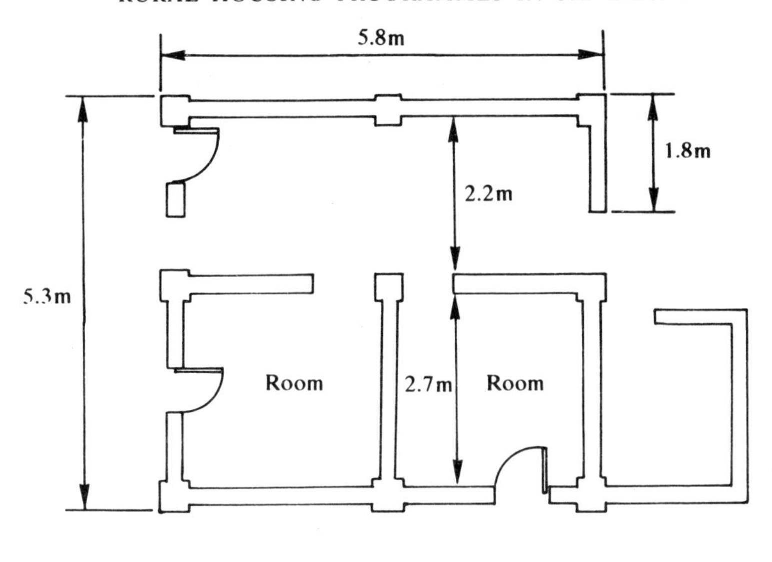

(a) Plan

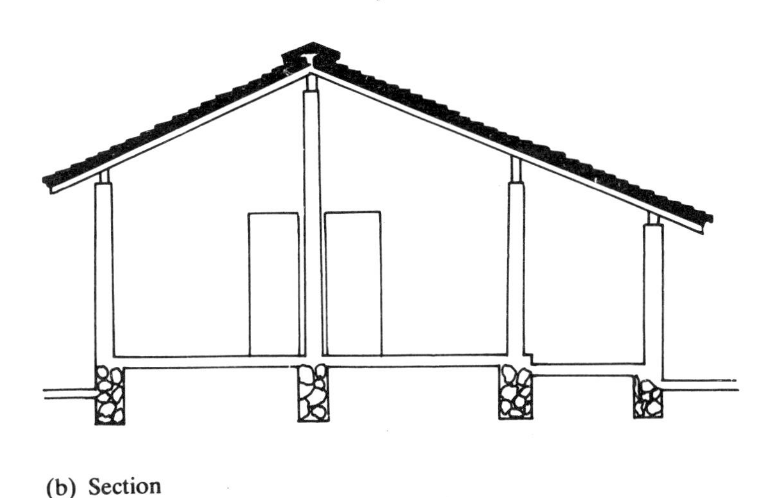

(b) Section

Figure 12.3 The L-4 house type plan introduced in Saumyagarn model village

by the male household head, yet he had apparently taken the requirements of the female household members into consideration. As seen in Figure 12.2, Plan 2, the space in the entire house is considerable and the kitchen is within the main premises, making it very convenient for the women. In all the cases studied of this type of housing, the recipients of housing loans found the amount given to be inadequate, and therefore borrowed from friends and relatives, except in one case, where the occupants could not find additional sources of funding.

A main problem faced by the recipients of housing loans in this type of scheme is they often ended up building a comfortably 'spacious' house using the loan money and additional funds but without proper toilets or wells. With the exception of one case, all other respondents had no proper toilets. Some walked up to 2 kilometres to obtain safe drinking water while others used common unprotected wells in the neighbourhood. In two cases, the houses did not have proper access roads; access was only across other peoples' property.

SERVICES, OWNERSHIP AND HOME-BASED EMPLOYMENT

Service facilities – such as schools, market, post offices, hospital and/or dispensary – in all these villages vary according to the type of housing scheme. In the Model Villages, these are all provided as part of the development package. The ASH and Reawakened Village schemes appear to depend on the facilities available in the traditional villages nearby. In the case of some schemes in the Kalutara district, householders (most often women with children) had to travel 5–8 kilometres to reach the schools and hospital.

In terms of ownership, except in the case of women heads of households, all others were owned by the male head of household. One of the houses in the sample chosen from the Reawakened Village Scheme belongs to the wife of the male household head because the house was constructed on the wife's property.

An analysis of the comments made by the women in these housing schemes regarding the possibilities of starting income-generating activities in their homes, revealed that most of them were of the opinion that small houses which are already limited in space for the normal living of the family cannot provide space for income-generating activities. Women in two households had found space in their compound for their small income-generating activities.

A comment made by most women was that they could hardly obtain any financial sources to start such income-generating ventures. Some were very keen to obtain such funding to begin home-based income-generating activities to supplement the family income and thereby overcome debt problems and improve their standard of living as far as possible. Poultry farming, bee keeping, lace making and coir making are some of the ventures mentioned. Funding is possible through the Thrift and Credit Co-operative Societies (TCCS) which have become the main financial institutions in the implementation of the Million Houses Programme. Although women are reported to be popular participants in the district-wide meetings of these societies (IRED 1987) and savings-account holders, a closer analysis would indicate that these women were largely those who were self-employed to begin with and had some reserves to invest. To begin such self-employed projects, therefore, women need an initial funding outlay and initiative. For this, the male members of the family need to give women adequate support both morally and physically.

DISCUSSION AND FUTURE RESEARCH

This study shows very clearly that beside the low involvement of women in Sri Lanka at the planning and designing phase of the rural housing programmes by the state, women and their housing requirements in terms of living space, physical comfort, space for income-generating activities, ownership and amenities have received low and woefully inadequate priority in the state housing programmes. The most recently formulated rural housing sub-programme, under which is placed the programme of Reawakened Villages, wherein houses are upgraded rather than rebuilt, marginally met some of the women's requirements. However, some amenities, such as toilet facilities and drinking-water supplies, seem to be a neglected aspect of this programme. Even under the ASH schemes, several instances where construction of toilets received low priority among the housing recipients could be seen.

The new rural housing sub-programme had 19 options known as the Housing Options and Loan Package. The main options are related to: upgrading an existing house; constructing a new house; a utilities package; and site and services provision. Quite often the recipients choose the first or second option and give lower priority to amenities. Once the householder obtains the largest loan (Rs16,000) meant for the site and the construction of the house, the monthly instalments become burdensome and the household never applies for additional funds to install amenities, thereby resulting in an unhygienic environment, particularly for women and children.

The need for assessment of housing-related requirements of women has been underscored at the international level by the United Nations and other agencies. This issue, however, has not been given formal attention by policy makers in Sri Lanka at the national level. The case for satisfying poor rural women's housing requirements is very pressing indeed.

Increasing attention, however, has been focused on the general situation of women in Sri Lanka in recent years as evidenced by the establishment of the Women's Bureau (1978) and the creation of the Ministry of Women's Affairs and Teaching Hospitals (1983) and the Ministry of Health and Women's Affairs (1989), to initiate and co-ordinate programmes to improve the position of women. Among the main functions of the Women's Bureau, advising the government in formulating policies and implementing programmes for the increased participation of women in national development and co-ordinating activities of institutions which strive to focus attention on women's causes, are of particular relevance here. The bureau has begun several income-generating projects for women in certain districts. It has also established centres for women to learn crafts like carpentry, masonry and bakery services which were previously male dominated. This endeavour has enabled several females to embark on self-employed projects and thereafter apply for housing loans to build houses that suit their requirements (Weerasinghe 1987; IRED 1987).

A few examples in certain districts (IRED 1987) indicate that self-employed projects improve women's access to housing. As mentioned earlier, access to credit, too, is possible through this pathway. To achieve this the low literacy level of women and lack of tradable skills are constraints that have to be eliminated.

A new and alternative path to rural development introduced through the Change Agents Programme has helped women considerably in certain districts to become more self-reliant and thereby manage self-employment projects effectively so that they can eventually satisfy their housing requirements. This approach has helped the women involved to break through the barriers of the traditional narrow attitudes which prevented them from becoming involved in progressive paths to development.

To obtain improved housing strategies for women in Sri Lanka, the immediate requirements are:

1 to formulate and implement housing policies and strategies at the national level that specifically recognise the role of women both as agents and beneficiaries in the housing programmes;
2 to devise programmes to improve infrastructure for water, energy supply, sanitation, transportation, education and employment so that women's burden of work is reduced, enabling them to contribute to the development process;
3 to design and construct on a self-help basis, houses specifically geared to meeting the physical requirements of women in maintaining family life and home-based income-generating activities;
4 to improve neighbourhood facilities to provide an adequate and safe environment;
5 to implement training and orientation services which include advice on various aspects of home ownership and maintenance;
6 to improve women's limited access to employment, especially in the rural areas;
7 to improve the educational standard of women to enable them to handle their basic needs more effectively.

13

WOMEN'S ROLES IN RURAL SRI LANKA

Anoja Wickramasinghe

INTRODUCTION

This study, carried out in villages in the dry zone of Sri Lanka, attempts to evaluate the multiple tasks undertaken by women in both small-scale farm operations and the domestic sphere. In this context, patterns of time allocation were studied over a period of one year in order to discern gender role differences.

NATIONAL EFFORTS TO INTEGRATE WOMEN INTO THE DEVELOPMENT PROCESS

Integration of women into the development process has recently become official policy in Sri Lanka. Much of the assistance towards development designed to promote the living standards of the rural poor has failed to help women, both in absolute terms and in relation to men. Therefore, the integration of women into development strategies both as agents and beneficiaries has become an economic imperative as well as a social equity goal. Yet the bottom-up development strategy, the Integrated Rural Development Programme (IRDP), which has been adopted for developing the rural areas of Sri Lanka since 1978 has not been able to obtain equal participation of men and women.

Although an attempt is being made at the national level by the Ministry of Plan Implementation and the Women's Bureau of Sri Lanka to promote women's participation in the process of development, this national effort has been largely ineffective in eradicating gender differences, especially in the rural sector. Instead of providing equal opportunities for both genders from the stage of planning development to the level of evaluation, a separate set of women's projects has been introduced. These projects aim at reducing poverty among rural women by providing assistance for income-generating activities, which tend to remain as a separate group of activities meant only for women. This is the situation noted in the package of IRDP which has been implemented so far in 12 administrative districts, i.e. Puttalam, Kurunegala, Kegalle, Nuwara-Eliya, Ratnapura, Monaragala, Hambantota, Matale, Gampaha, Kalutara, Badulla and Matara.

Isolation of women's projects from the rest of the constituents in the IRDP package, automatically limits the proper participation of women in the strategy of rural development, which aims at promoting the physical infrastructure, the economy and human resources. On the other hand, the plans formulated and implemented at the national level often suffer from lack of baseline information, and so are unable to meet the needs of rural women satisfactorily. However, failure to recognise the vital role played by women in developing rural areas, weakens the proper integration of women into development. The recognition of the multiple tasks played by rural women in the household unit, in their society and in the economy is a prerequisite for the formulation and implementation of plans for sustainable rural development.

WOMEN IN THE AGRICULTURAL LABOUR FORCE

In Sri Lanka, as is the case in most countries, the economically active group is male dominated. According to the Census of Population in 1981, the proportion of economically active men was 74.5 per cent. The female labour force was only 25.5 per cent of the total labour force but this was exclusive of women who were engaged in production activities mainly during the peak agricultural seasons. This gender disparity in labour-force participation is a well-marked feature in all age categories (Figure 13.1), but between 1963 and 1981 there was a slow expansion in the participation of women in the labour force. The highest rate of women's participation in the labour force is in the age group 20 to 24 years and it remains almost constant until 34 years, and then decreases rapidly. In Sri Lanka, this reduction is often associated with the increasing demands of women's reproductive role. On the other hand, the highest rate of male participation occurs between the ages of 35 and 39 years and then decreases slowly until 49 years. Thereafter a rapid reduction in the participation rate is noted.

The rural share of the labour force is 78 per cent in Sri Lanka, mainly due to the confinement of 78 per cent of the population to the rural areas. Just as in any other production sector, among agricultural sectors too the gender role difference is well noted. The proportionate share of women in agricultural sectors varies spatially and seasonally. One of the notable features of agriculturally based production activities is the absorption of a comparatively high proportion of women workers. Among women, 52.4 per cent work in agriculture, whereas only 42.8 per cent of the male labour force works in the agricultural sector.

A well-marked disparity in the gender distribution of labour among

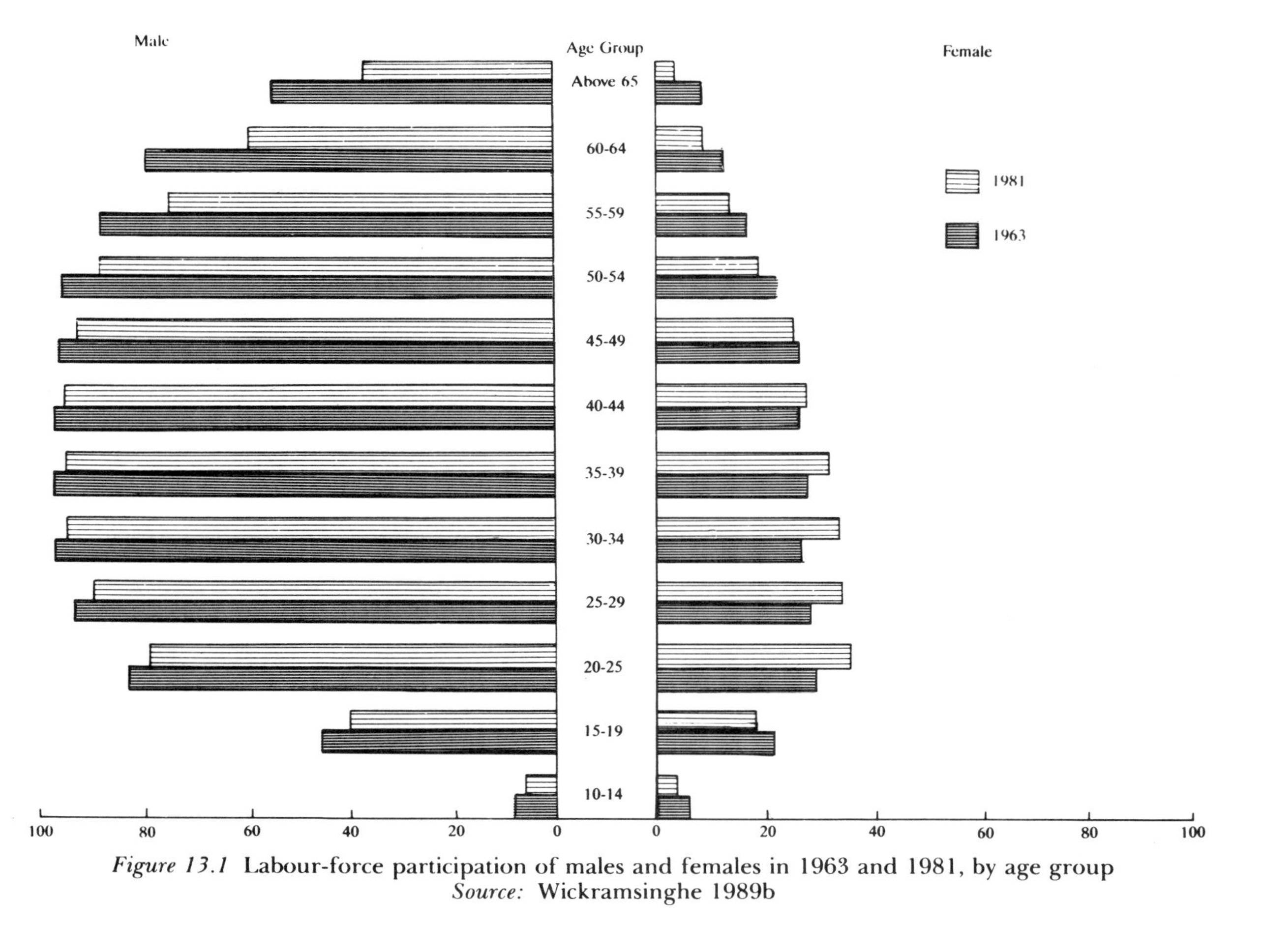

Figure 13.1 Labour-force participation of males and females in 1963 and 1981, by age group
Source: Wickramsinghe 1989b

Table 13.1 Division of female and male labour in agricultural and fishing activities

Category	*Female*	*Male*	*Total*
Paddy	17.1	46.0	38.8
Rubber and tea estate	67.8	19.9	31.7
Coconut estate	1.2	1.6	1.5
Other cultivation	10.1	17.7	15.8
Other agricultural work	3.9	9.8	8.3
Fishing	0.0	5.1	3.8
Total	100.0[a]	100.0[a]	100.0[a]

Source: Adapted from Census of Population and Housing 1981
Note: [a] Some figures are rounded

the sub-categories of agriculture is noteworthy. Here, the influence of the high rate of female participation in the plantation sector on the overall female participation in agriculture can be observed. Almost 68 per cent of the females in the agricultural sector work in plantations, in intensive activities such as tea picking, weeding, and rubber tapping (Table 13.1). Men's share in this sector is comparatively low, accounting for only 20 per cent of the total male workforce in agriculture, and is mainly involved in heavy labour and supervisory work (see Figure 13.2 for details).

As shown in Table 13.1 the proportionate share of labour in paddy cultivation is high having 38.8 per cent of the total agricultural labour force. This accounts for 46 per cent of the males and 17.1 per cent of the females in all agricultural sectors. However, if one deconstructs paddy cultivation in order to identify the division of female labour according to employment status, a rather different picture is highlighted (see Table 13.2). Paddy cultivation which is mainly meant for subsistence is confined to small-scale farming and it is mainly an owner operated system. The proportionate share of female labour is low in comparison with the male's share, but the percentage of women working on their own farms is higher than in the other categories at 63 per cent. Moreover, almost another 30 per cent of the women work as unpaid family workers in paddy cultivation while male labour accounts for only 6 per cent.

In addition to their engagement in paddy cultivation women play a prominent role in small-scale farm operations. These include the activities which deal with the production of dryland crops in the dry zone, and vegetables in the hill country and also the management of homesteads to produce a part of the daily food requirement of the family. In most of these productive activities, women work as unpaid family workers to produce a large proportion of the food requirements of the family and to subsidise their family income.

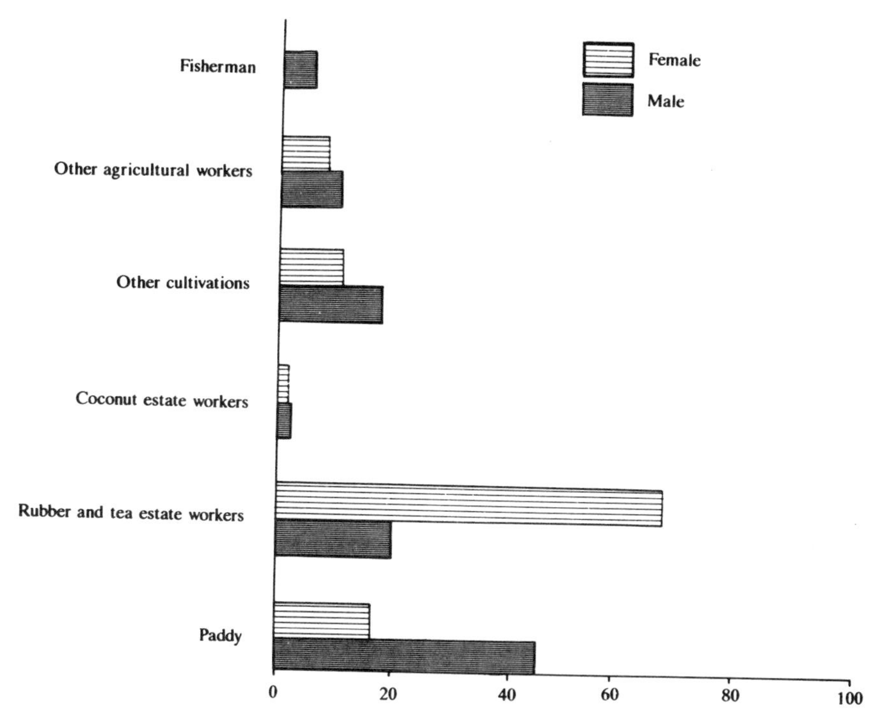

Figure 13.2 Distribution of male and female labour in the agricultural and fishing sector in 1981
Source: Wickramsinghe 1989b

Table 13.2 Division of labour in paddy cultivation

Category	*Female*	*Male*	*Total*
Own account workers	63.1	84.0	81.8
Paid employees	7.2	9.6	9.3
Unpaid family workers	29.7	6.4	8.9
Total	100.0	100.0	100.0

Source: Adapted from Census of Population and Housing 1981

The gender role difference in agriculture

Gender role difference is a well-marked feature of both domestic and production activities in the agricultural sphere. This is significant in subsistence agriculture where agricultural operations are carried out by family labour and the adoption of new technologies is rare. In the present study, which was carried out in the area of Mauswewa village (Figure 13.3), the high rate of women's involvement in farm activities in

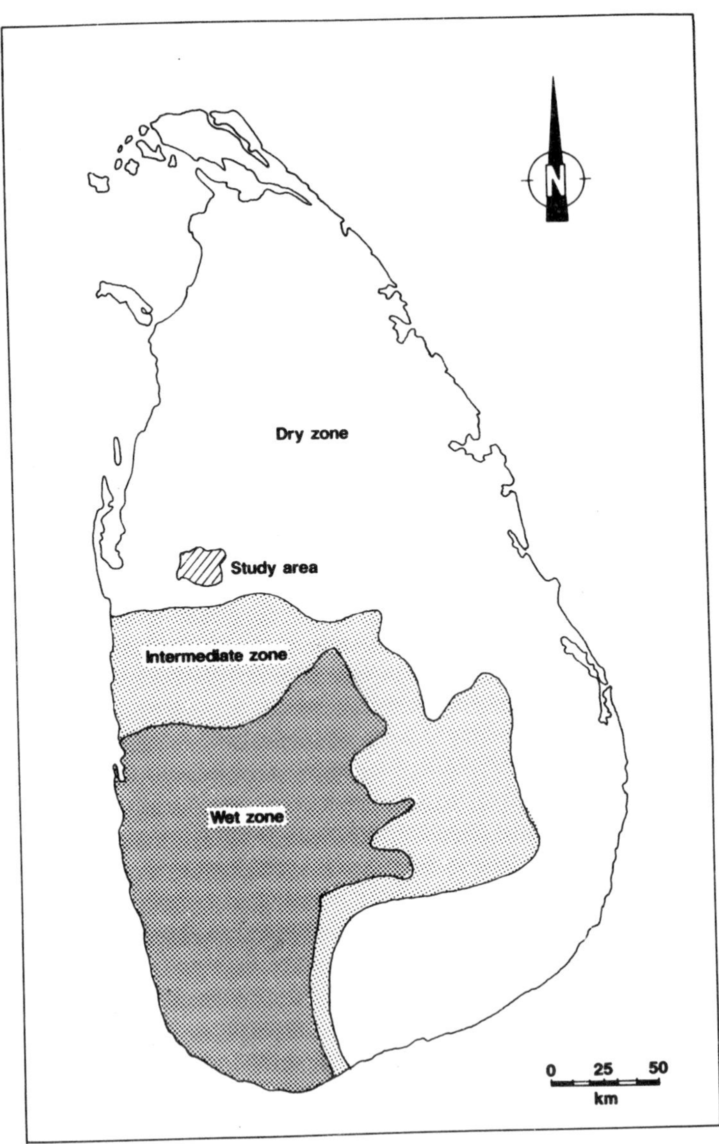

Figure 13.3 The major ecological zones of Sri Lanka and location of the study area

both lowland paddy cultivation and highland cultivation is a prominent feature in small-scale farm operations. One of the features noted here in the agricultural sphere is the dominance of women in time-consuming and labour-intensive tasks while heavy work such as ploughing of fields using draught animals, the sowing of paddy, the transport of harvested produce, threshing etc. is attended to by males. This gender-based division of labour in paddy farming is traditional, and is partly due to the acceptance in this culture of the female nature of time-consuming work such as transplanting and weeding, while heavy labour in water-logged fields is considered men's work.

However, in dry land farming, women's control over production activities is much greater than that in paddy farming (Table 13.3), partly due to the nature of the farming system itself. In addition, in the highlands, a few varieties of pulses, legumes, tubers, and vegetables are grown, either in rotation or in mixtures during the rainy season, and therefore labour requirements are much greater. Sequential planting, sowing, weeding and harvesting are the major activities that take place throughout this cultivation period, and as a consequence, unlike in paddy farming, the gender division of labour in farm activities is less significant. Women attend to a range of activities such as clearing the land, weeding, burning of shrubs, sowing, harvesting, processing and winnowing. Almost 65 per cent of production activities in dryland farming are done by women while in paddy farming they carry out only 30 per cent of the tasks.

Table 13.3 Sexual division of labour in agricultural activities

	Paddy cultivation %		*Dryland cultivation* %	
Activity	*Female*	*Male*	*Female*	*Male*
Field preparation	0.0	100.0	50.0	50.0
Sowing	0.0	100.0	60.0	40.0
Weeding	90.0	10.0	90.0	10.0
Harvesting	60.0	40.0	80.0	20.0
Threshing	0.0	100.0	30.0	70.0
Winnowing/cleaning	80.0	20.0	80.0	20.0

Source: Author's fieldwork

Traditionally, women, as wives and mothers, are expected to see to the well-being of their family. With increasing household requirements and hardship in rural society, their role in domestic food production as unpaid workers has become prominent. Males are expected to migrate to earn cash and to do the heavy work, while women, in addition to farming activities, are confined to bringing up children and processing and preparing food. The multiplicity of women's responsibilities in rural society has become a prominent feature.

Women's tasks in agricultural crop production

With the introduction of high-yielding varieties of crops to dry-zone farming, particularly the short durational varieties which are largely meant for the market, the involvement of women in agricultural production has tremendously increased. This fact contradicts Boserup's (1970) work in Africa where, according to her observations, men were gradually taking more responsibility in agriculture (especially in the growing of cash crops) and at the same time women's workload and status in general appeared to have deteriorated. However, in Sri Lanka, the increasing need to produce food for family consumption and for the market, often forces women to invest their time and energy in agricultural activities, especially in small-scale crop production.

In the dry zone of Sri Lanka, the organisation of agricultural production is largely controlled by the agro-climatic situation in which major crop production is confined to the rainy season. With the introduction of market-oriented cash crops in place of subsistence food crops, in association with the confinement of dryland farming into permanent blocks, women have taken on greater responsibilities for producing a range of crops in the highlands. With the increasing requirements for women's labour input in agriculture, particularly during the peak seasons, women perform as waged labourers, unpaid family workers and workers in reciprocal exchange of labour.

Although the last two forms of labour are quite common, the supply of waged labour is determined by factors other than the availability of work and labour demand. The status of the family and its wealth, and the areal extent of their own agricultural land are the influencing factors which determine the engagement of women in the waged labour force. Women of low-income families often seek waged work to earn a cash income.

However, women often get paid at least 10 to 20 per cent less than men even for equal amounts of work. Women's role in agriculture as unpaid family workers is a common phenomenon in remote dry-zone villages. Lending their own labour on an exchange basis for collective work is a traditional established system and is still prevalent among farm families. It is advantageous in activities such as transplanting and harvesting where work has to be completed within a limited time.

Seasonal variations in time-use patterns

One of the features noted in the study area is the restriction to one prominent crop during the *maha* season, between October and February, and a second minor crop during the *yala* (*vala*) season between April and August. This second crop is not widespread, but is found in some areas when and if the south-western monsoon rain occurs (Figure 13.4). Seasonal changes in labour requirement and the availability of agricultural work have a direct impact on the pattern of women's involvement in both agricultural and domestic activities. During the slack seasons

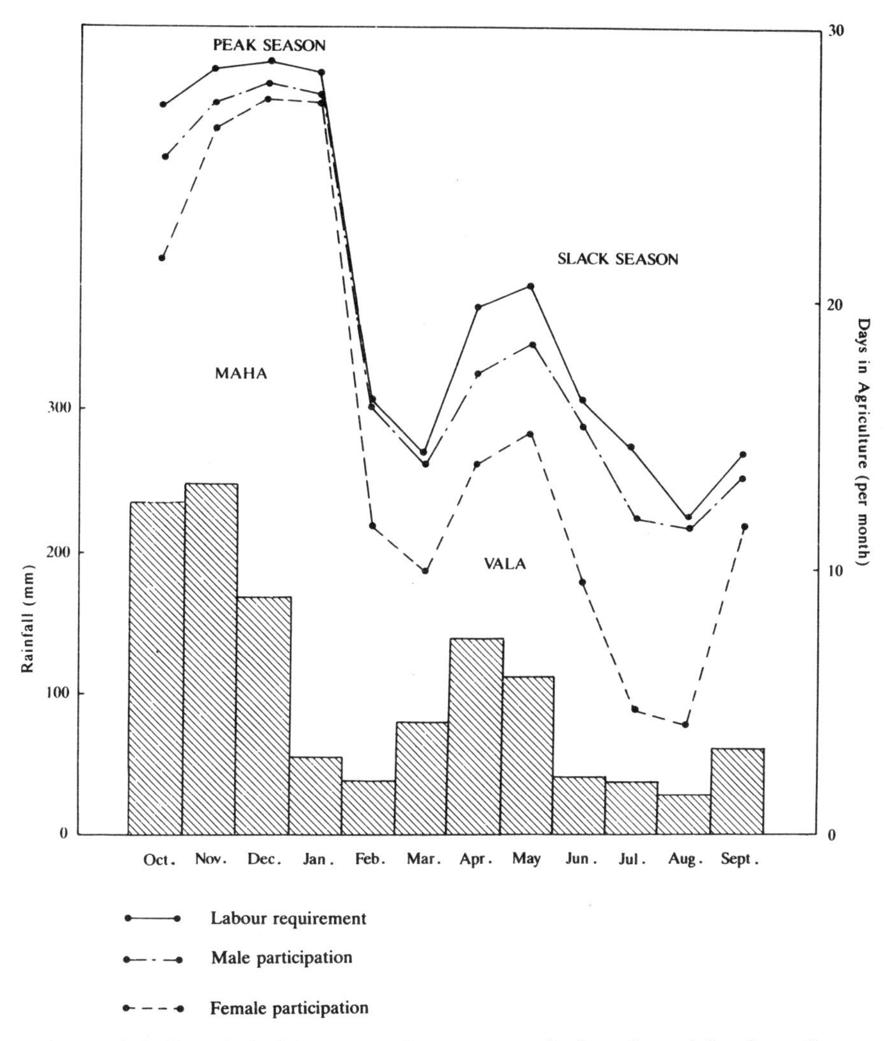

Figure 13.4 Trends in labour requirements, agricultural participation of men and women and the pattern of annual rainfall distribution in the study area

women's role in crop production is comparatively less than that of men and than in the peak season, due to the low labour requirement. In such situations, male members of the family spend their time in the field while women spend more time on home-based activities.

In addition to the gender disparity in time use, another feature embedded in the annual time-use pattern is the seasonal difference in dividing time between domestic and production tasks (Figure 13.5). Activities contributing to reducing household food costs and to increasing

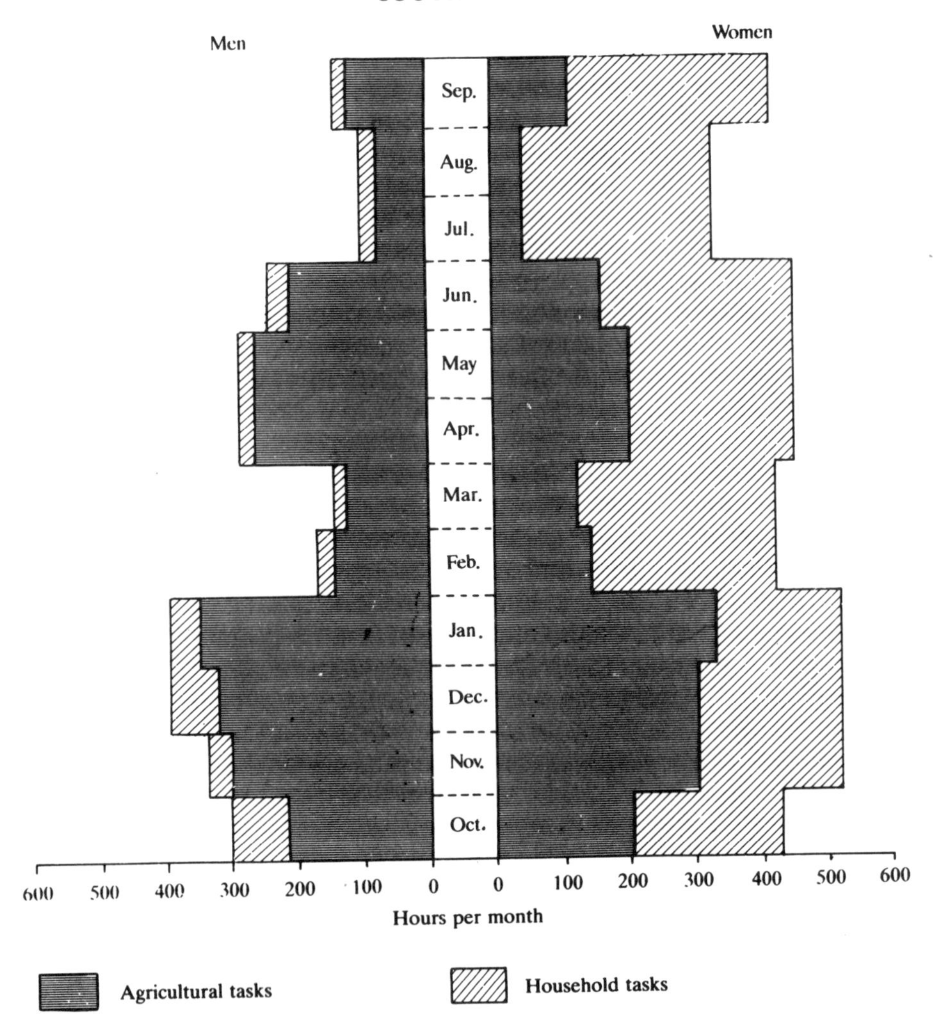

Figure 13.5 Gender disparity in the seasonal and annual time-use pattern (average number of hours per month)

household food security are a part of women's domestic burden. In this peasant society, the time allocated by farm women to domestic tasks often increases during the agricultural slack seasons. This situation is reversed during the peak seasons in agriculture when extra time is needed for the processing and preserving of food to be used during periods of food scarcity or sold in hard times. The major home-based processing activities are the drying, cleaning and storing of rice, cereals and chillies and the extraction of medicinal and consumable oils.

The allocation of time by women to agricultural production is only

Table 13.4 Time spent on agricultural and domestic activities by men and women in the dry zone (hours per month)

Activity	*Agric. peak season* Female	*Agric. peak season* Male	*Agric. slack season* Female	*Agric. slack season* Male
Production activities (Total)	299	298	235	245
Of which:				
Household activities	189	168	160	100
Wage/exchange labour	110	130	75	145
Housework (Cleaning, cooking)	199	90	220	60
Fetching water/collecting wood	50	30	60	30
Social/religious	12	08	15	15
Leisure/sleep	160	294	190	370
Total working hours	560	426	530	350

Source: Author's fieldwork

slightly less than that of men during the slack season (Table 13.4). Yet, the total number of hours devoted by women to work both in agriculture and in the house is considerably higher than that of men. Therefore, if one considers the total number of hours spent on work, then the woman's contribution to the family in the form of production and family expenditure savings is higher. Moreover, the inadequacy of the time available to women for leisure is bound to exhaust them.

Fieldwork demonstrates that if the total workload of women is taken into consideration and evaluated correctly, then the contribution of the woman to a family unit is at least 32 per cent greater than that of the man. Although men appear to have a heavy workload, women bear the burden of attending to a range of farm activities and often work longer hours to complete the work in the field and in the household, sacrificing their leisure and personal requirements. Women are more important in crop production in the dry zone where the multiple activities in agriculture are clustered into one season and hired labour is difficult to obtain and the use of machinery for cultivation practices is either limited or beyond the capacity of small-scale crop producers. Nevertheless, the vital role played by women in crop production is not adequately recognised.

WOMEN IN THE HOUSEHOLD UNIT

The activity-based division of female labour into reproductive and productive is rather vague, firstly, because crop production for family consumption is taken as part of reproduction, and secondly, because women often attend to both productive and reproductive activities simultaneously. For example, the collection of fuel is often done while clearing land for cultivation and the firewood is carried home after completing the day's work in the fields. Especially during the peak seasons in agriculture, many women, while engaged in crop production, try to do domestic work as well. During such periods, they change their activities more often than during the agricultural slack seasons. Often mothers of small children take their children to the fields, and nursing, bathing and feeding are done while attending to farm work. Similarly, production of crops for the family plot is neither considered as a gain nor as a way of saving family food expenditure.

Gender role difference in household tasks reveals a greater multiplicity of women's roles than of men's (Table 13.5). This is not only in the number of normal responsibilities undertaken by women, but also in taking family decisions and handling the family income. In addition, maintaining the family's nutritional status, health, and children's education are a part of women's responsibilities.

It was noted in this study, that in an instance like a wife's sickness only 8 per cent of men could cook. Very few men, about 5 per cent, help their wives in the home, but even among these few, child care, the collection of fuel, and lighting the fire in the hearth are the most

Table 13.5 Gender differences in participation in household activities (male and female shares are calculated as %, based on the total number of hours spent weekly on each activity)

Activity	*Female*	*Male*	*Total*
Food processing and preparation	92	08	100
Winnowing and parboiling rice	100	00	100
Fetching water	98	02	100
Post-harvest grain storage	30	70	100
Collection of firewood	65	35	100
Preserving of food for later use	80	20	100
Keeping house and yard	95	05	100
Producing vegetables, tubers and fruit for family consumption	80	20	100
Raising of children	90	10	100
Bathing of children	80	20	100
Attending to sick in the family	85	15	100
Guiding children's education	97	03	100
Washing clothes	100	00	100
Caring for the elderly	90	10	100
Participation in village ceremonies	45	55	100
Participation in village social activity	10	90	100
Participation in village community development	05	95	100
Rearing animals (goats, cows, buffaloes)	50	50	100
Milking cows	100	00	100

Source: Author's fieldwork

willingly undertaken tasks. Quite commonly, the burden of household work is shared by the children. The girls of the family often substitute for their mothers in looking after young children, carrying water and firewood and helping in the kitchen, while boys lessen the burden on the adults by helping in the fields and feeding livestock. Such substitution is often required, particularly during the peak season in agriculture when adults work longer hours in the fields to increase both income and crop output.

The aspirations of women in the study area are clearly directed towards the well-being of their family. A better standard of living, housing, clothing, and children's education, plus a stable income, employment opportunities for the next generation and land for crop production are the priorities mentioned by almost 173 housewives out of the 200 families living in the study area. Individual goals mentioned by women are rare and, in most cases, the physical burden of their multiple roles is taken as a part of their responsibility. Even better food for their own satisfaction is of little importance to them.

A gender difference in decision making is observed in both domestic and non-domestic spheres. Most public performances at social organisations are carried out by men, while female participation in village organisations is limited to 35 per cent, and this too is mainly as a member of a religious or death assistance society. Hence it is not women's subordination which permits men to play a leading role in social matters, but rather lack of time and the difficulty of being present at meetings for organising social welfare and development programmes. However, in 92 per cent of the cases studied, household heads tolerate and appreciate the expression of views on community matters by women, but, in general, they all accept the practical difficulties which prevent women's active participation in community-level organisations due to the extremely heavy burden of domestic work. In discussing matters relating to the community and the financial situation of families at village organisations, the ideas and consent of female partners are taken into consideration.

However, the general situation noted in this study is quite different from that noted by Leach (1960) in the dry zone, where a Sinhala husband and wife may work together in the fields, but there are no other occasions upon which they can be seen together in public with propriety. The purchasing of clothes for special occasions, attending ceremonies and religious festivals, visiting patients, and visiting friends and relatives on special occasions by husband and wife together were observed in this study area. Similarly, unlike the case noted by Wickramasekara in 1977, in which he identified the leading role of the household head, the husband, in dividing the work among the family, in this present study, both husband and wife discussed the labour requirements and divided the day's work. In very many cases, it is the wife who decides what should be done during the day by individual family members.

As observed in an earlier study (Wickramasinghe 1989a), the leading role played by women in decision making within the family unit is bound to give better status to women in the domestic sphere. Almost 94 per cent of the husbands, the household heads, agree that women's services are indispensable. Finally, matters relating to the preparation of food for children and the elderly, seeing to children's education, family health and nutritional status, family planning and financial matters are better handled by women than men. However, as both men and women agree, the husband, the head, formally bears the responsibilities for the security of the family. The common consciousness of the importance of the women's role in the family unit is rewarding. Excluding 12 per cent of the families studied where husbands were addicted to liquor or treated their wives as inefficient, most women are accepted as better managers than men in regulating family expenditure and handling income. In general, in handling financial matters, as in any other matter pertaining to the family, some mutual consideration between men and women exists. It is very much so in the families in which the man's individual interests do not extend beyond purchasing items needed for chewing betelnut.

Selling of excess agricultural products at the outside market, which is located about 10 kilometres away, is often done by the men. Nevertheless, the men give the cash obtained from such sales to their wives on their return from the market, and whenever possible, save a little for the future after mutual discussion. Decisions on major expenditures, other than foodstuffs purchased from the outside market, are made jointly.

However, financial matters related to credit and insurance are handled by men. The formal channels for obtaining credit are open to men due to their ownership of property, particularly land. Therefore, as noted by Palmer (1978) access to the resources which are needed in the application of agricultural technologies is in the hands of the men. Under these circumstances, provision of assistance or credit for the adoption of new agricultural technologies is mainly under the control of men.

Although all the women in the studied villages are engaged in agricultural activities either directly or indirectly, only 3 per cent are given opportunities for further training. This training is limited to dairying which is a traditional activity of the area. However, if knowledge of desirable low-cost agricultural technologies was widely spread among women, perhaps by introducing a mobile service, then women's self-confidence in farming would be improved.

If one examines the situation which prevailed in rural areas about two generations ago, it appears that society's expectations of women and the multiplicity of their responsibilities was less than at present. The expansion of families, scarcity of resources, particularly the land on which their families survive, and the inadequacy of crop production for a family's survival have a direct impact on women's tasks forcing them to work equally with the male partner in productive activities. Many complexities and responsibilities have been added to their traditional role of caring for children and elders, processing and preparing food and keeping house. On the other hand, the help that women have had from their children is tremendously reduced in rural society because of the increasing rates of school attendance. Instead of the assistance that women have had from their children, particularly from female children, in their domestic work, and in rearing animals, it has become a responsibility of the mother to see to the children's educational achievements. The financial hardship and the extra workload born by rural women are heavy, and consequently she is bound to involve herself in a range of non-domestic activities to meet increasing family food requirements, as well as to produce a marketable excess for cash or occasionally engaging in paid work, if possible.

CONCLUSION

The main observations of this study help to highlight a number of complexities and difficulties which could practically arise in promoting women's economic and social status in rural areas.

Within small-scale agricultural operations in the dry zone, women's production activities are of vital importance, not only for better crop production but also for the survival of the peasant community. If one examines the variations existing in women's role in agricultural operations across the country, women's agricultural involvement in dry-zone areas is notable.

The division of labour in both agricultural production activities and in the household is partly a result of the acceptance of women's domesticity and femininity as a traditional norm. The double burden of work often discourages women from utilising the opportunities available for acquiring a vocational training, holding leading positions in the community and engaging in income earning away from the family.

In Sri Lanka the goal of equal opportunities and equal status are largely submerged in the aim of improving the livelihood of the society in general. Rural women in traditional society are obliged to work for the family well-being. This is not merely society's expectation but an attitude of rural women themselves. As expressed by almost 90 per cent of the wives in the study area, motherhood and affection bound within the unit of a household, 'a family', are prized by rural women. Often the mother's efficiency in handling those household matters which are traditionally women's responsibility, gives her greater control and autonomy in the domestic sphere.

However, to improve the living conditions of rural women and integrate them into the development process, it is necessary to widen their training and employment opportunities, increase incomes and relieve women of the extremely heavy burden forced on them in the domestic sphere. The promoting of women's productive activities by the provision of income-generating avenues, assistance, training and exploitation of marketing facilities is convincing. But, as previously shown by the author (1987a and 1987b), in examining the projects introduced under the Integrated Rural Development Project (IRDP) in the Kegalle district, the effectiveness and the sustainability of such projects to meet the broader objectives of improving the status of rural women cannot be assured. However, in the dry zone the prospects for introducing agro-based cottage industries are greater where there is diversity in crop production and seasonal variation in cropping exists. It is important to stimulate occupations preferred by women, especially in oil crushing, manufacturing dairy products, grain milling and food processing, but, as mentioned earlier, desirable training, assistance, motivation and marketing avenues are prerequisites.

The main problems which arise in promoting women's employment are related to time. At present, women tend to spend much time in the fields during the cultivating season. The additional time needed for agricultural activities is gained at the expense of leisure and sleep, and by reducing the time spent on caring for children and domestic work. The time available to the dry zone women in the study area for sleep is extremely low and unsatisfactory, particularly for women with small children and infants. On the other hand, it is not possible to find substitutes for women in the domestic sphere. Most families cannot afford hired help and even if they could, Sri Lankan employees would not be prepared to work long hours for another family's well-being. If a woman gives secondary importance to her domestic role, as a mother or a wife, then the unity, the happiness and the harmony of family life will deteriorate. Under these circumstances, the goal of equal opportunities and a satisfactory standard of living can only be achieved by increasing the work share of the male partners in the domestic sphere and reducing time constraints for women. Sharing of housework by the male partner would certainly reduce the time spent on domestic activities by women by almost 40 per cent. Changes in society's expectations and attitudes might increase men's involvement in housework. However, it is difficult to expect this to occur quickly or among older people, and therefore, knowledge of domestic activities should be provided equally to both male and female children from the beginning of their school careers. It is also necessary to formulate rural societies which promote collaborative work.

Provision of child-care facilities would relieve women of both physical and mental stress. The labour necessary for such purposes could be obtained from the village community itself either on a rotational basis or on a wage basis. Accessible water, rural electrification and simple technological improvements in the processing and preparation of food in the home would also be a great help. However, purchasing and maintaining such facilities is costly, and so they cannot be provided for low-income communities where employment is irregular and income inadequate to meet the bare necessities of life. Even payments for electricity supplies are not within the means of 82 per cent of the villagers. Therefore, technological improvements for the rural community must follow improvements in economic and financial security.

It will take time as is noted by Whyte and Whyte (1982) to eradicate the acceptance of a secondary, submissive role for rural females, and the automatic allocation to males of a greater right to education. This situation often prevents women from holding leadership positions and taking initiatives in village-level activities. Gender role differences can be reduced by removing the social stigma of domestic work. One way of achieving this goal is through mutual understanding. Another way is by increasing the self-confidence and self-reliance of women themselves. An economic evaluation of women's tasks both in production and welfare activities would encourage society to give due recognition to women's work. If such efforts are made then the equity goal and changes in social attitudes will be achieved gradually.

Part IV

SOUTH-EAST ASIA AND OCEANIA

INTRODUCTION

Gender and geography in South-East Asia and Oceania

A regional overview

Janet Momsen with Nancy Davis Lewis

There are very few geographical publications on women and development related to this region (Chiang 1988). However, despite quantitative limitations, it is possible to identify a distinct focus. Most researchers appear to be looking at the confrontation between modernisation, in all its ramifications, and traditional livelihood systems. This confrontation can be seen to have both negative and positive aspects: with proletarianisation leading to both the undermining of women's position because of Western influence (Buang Chapter 15; Fahey 1986) or to greater independence for women because of increased income-earning ability (Singhanetra-Renard 1987); better living standards result from access to new material goods (Fairbairn-Dunlop 1991) but lead to concomitant declining standards of production in traditional women's craft work (*ibid.*) and the development of a consumer society for such non-essential, non-traditional items such as cosmetics (McGee 1987).

These studies fall into two groups: those on migration and those on changing village life. These topics are not mutually exclusive. John Connell (1984) in his review article on women's migration and development in the south Pacific notes that data on this topic is very limited. Nora Chiang (1984) described the migration of rural women to Taipei and suggests that psychological and social aims may be as important as economic forces in the decision to move from the village to the city. Fahey (1979) looked at labour migration to Bulolo in the Highlands of Papua New Guinea to work in mines and on plantations and found that changes in the labour process in the initial stages of capitalist penetration are influenced by a pre-existing sexual division of labour and by the expectations of the colonisers. Both Larner (1990) in relation to Samoan migrants in New Zealand and Buang (Chapter 15) studying Malay women moving from villages to work in export-zone factories found that these women migrants are seen as a source of financial assistance by their natal families although the women themselves have many social problems in the adjustment to alien 'Western' ways. Peggy Fairbairn-Dunlop (Ch. 16 and 1991) has noted the dependency of villagers on western Samoan relatives in New Zealand when funds are needed for village ceremonies. Consequently, status in the village has come to be related to the financial success and generosity of emigrants. On the other hand, Singhanetra-Renard (1987) provides a positive view of migration in northern Thailand. The independent circular migration of both men and women provides a vital contribution to rural livelihood strategies. She has also linked this spatial mobility to declining fertility levels (Singhanetra-Renard 1984) while Young and Salih (1987) relate changing Malay family patterns to structural change, involving migration, in the national economy. Housing is also affected by spatial mobility and the loosening of familial ties (Ahmad Chapter 14; Buang Chapter 15; Bryant 1991).

These broader changes are also considered in terms of gender differences by several authors. Stephanie Fahey (1986) examined the relationship between 'subsistence' production, simple commodity production and wage labour in a peri-urban village just north of Madang in Papua New Guinea. She argues that the village as a social group is dependent on wage labour for its reproduction and hence is proletarianised. As part of this process, married women in the village have become doubly subordinated: to capital and to men. Rebecca Elmhirst (1989) looked at the relationship between capitalism and gender inequalities in Indonesia while David Preston (1986 and 1989) has considered gender aspects of the expansion of uncultivated land in south-east Asian villages with a high proportion of the working-age population going to work in the city, as has Sando (1981) for Taiwan. Other geographers have looked at the impact of technological change on specific aspects of life in the region. Kim DesRochers (1990) has studied the effects of cultural, social and technological change on women's use of the near-shore zone for fishing on Kosrae Island in the Federated States of Micronesia. Pam Thomas (1986) has addressed the long-term impact of a western Samoan women's health-care development project started in 1923 and the problems caused by its Eurocentric orientation. She has also considered the role of Pacific women in income generation (Thomas and Noumea 1983) and stresses the importance of information and communication.

Overall, these papers present a vision of rapid change and a traditional lifestyle under threat. However, in this dynamic situation there appears to be growing pressure on women as the keepers of traditional culture and ideology.

Feminist geography in Taiwan and Hong Kong

Nora Chiang

There are about as many geographers who belong to Universities in Hong Kong as in Taiwan. However, the number of female faculty who have earned Ph.D. degrees is higher in Taiwan, with one department having a female chair. Despite this, few of the women geography faculty in Taiwan are in social sciences and so their impact on the curriculum is small. Although feminist geography has been introduced to Taiwan by international scholars through lectures, publications and newsletters, there has been little follow-up work. So far there has been one M.A. thesis completed on working women in Taiwan and one research project, undertaken by myself, on factory women.

My efforts to pursue feminist geography have been carried out in various ways. My study of female migration in Taiwan complemented previous migration studies which were either on male migration or ignored gender differences. However, this effort to correct gender bias is not limited to geography alone. In 1985 I co-ordinated the conference on 'The role of women in the development process in Taiwan', and I founded the Women's Research Programme (WRP) in the same year. The annotated bibliography compiled by myself and my assistant has been expanded into a second and third edition. The WRP, now in its seventh year, has enlarged its objectives of promoting women's studies, to providing resources in the form of materials and information for the academic and non-academic community. Its research activities have included public lectures, luncheon seminars and conferences, as well as publication of a bulletin, a journal, monographs and occasional papers in both English and Cantonese. The growth is due to public demand, mainly in the academic community, and the dedicated efforts of its members. Many courses are taught at various universities in Taiwan with a feminist perspective, including the programme's course on gender relations which had an enrolment of over 400 in 1992.

In Hong Kong, the Gender Research Programme represents the counterpart of the WRP in Taiwan, within a much smaller academic community. Without any participation from geographers, the programme includes a much wider range of disciplines than in Taiwan. It is established as one of the seven strategic programmes of the Hong Kong Institute of Asia–Pacific Studies. It was started in the same year as the WRP, 1985. Its members took on the task of printing a newsletter and have established a data base and documentation unit on women in Hong Kong. With its international location, the Gender Research

Programme has organised two conferences which have been attended by mainland Chinese, Hong Kong and Taiwan scholars. This has been very important as it is still not possible for scholars of mainland China and Taiwan to visit each other on official occasions.

A much smaller number of courses on feminism exists in Hong Kong, probably due to less flexibility in the curriculum. When courses are offered, large numbers of students enrol. When I visited Hong Kong during the academic year 1990–1, as an Elizabeth Luce Moore scholar, I suggested two changes in the curriculum of the Department of Geography at the Chinese University of Hong Kong. The course 'Man-environment systems' dropped the word 'man' with the consent of the lecturer and was renamed 'Environmental management'. The course on frontiers in geography added a section on feminist geography. Working with an all-male faculty, I did at times point out the absence of other female faculty members. To my astonishment, many departments at the Chinese University of Hong Kong do not have any women on the faculty. On the contrary, in Taiwan, out of six departments in the Faculty of Arts, four chairs are female. Only one department in the Faculty of Science has a woman departmental chair.

In the Department of Geography in which I teach, further attempts are being made to introduce feminist geography at the undergraduate level. In the team-taught course 'Gender relations', I am giving a lecture on women, space and the environment. In an effort to expand the curriculum into an inclusive one, I introduced exercises on feminist geography in the first-year human geography course, and included two lectures on gender topics in the course on social geography. Encouraging my students to use the library at the WRP is necessary in order to raise the awareness of students towards feminist issues. These steps are taken rather slowly since the structure of the curriculum is still oriented towards physical geography and technical skills. In an elective course on environmental ethics, a section on ecofeminism was introduced by a visiting professor.

If feminist geography aims to change people's way of thinking, it still has a long way to go in Hong Kong and Taiwan. Due to the composition of faculty members and students, the structure of courses and factors such as job markets, changes only come slowly. The Women's Research Programme which stresses inclusiveness and connectedness, should be a major camp in which female spirituality is found in the academic world. Needless to say, women's studies represents the women's movement into academe. With its dominance in thinking and action, it does provide a place for all disciplines to co-exist. It is undoubtedly one of the most successful programmes in the university and remains the only one of its kind in Taiwan.

14

GENDER AND THE QUALITY OF LIFE OF HOUSEHOLDS IN RAFT-HOUSES, TEMERLOH, PAHANG, PENINSULAR MALAYSIA

Asmah Ahmad

INTRODUCTION

Structural changes in the Malaysian economy, particularly since the 1970s, have resulted in the increased involvement of women in wage work. The shift in emphasis from primary industries to secondary and tertiary industries has consequently released a major portion of the workforce from subsistence agriculture to wage labour. With the advance of urbanisation (from 26 per cent in 1957 to 37 per cent in 1980) and industrialisation, Malaysian women too find themselves seeking wage labour.

Industrialisation and the pace of urbanisation in certain areas have provided a focal point for youth inmigration and the subsequent problem of shelter. Acute housing shortages, high house rental and poverty are often quoted as factors leading to squatting. In Kuala Lumpur, the capital city, for example, most studies tend to estimate the city's squatter population at between one-fifth and one-quarter of its total population. This means that out of the city's total population of 919,610 in 1980, between 185,000 and 230,000 inhabitants were squatters.

Whilst most shanty towns in Malaysia are found on land occupying low-lying riparian areas, disused mining sites and land adjoining railway lines, this chapter attempts to highlight a less common form of urban squatting, that is, living afloat on raft-houses. Although such a living environment is unusual in Peninsular Malaysia, it is without doubt an example of human ingenuity in overcoming the urban housing problem and the fight for survival both physically and economically.

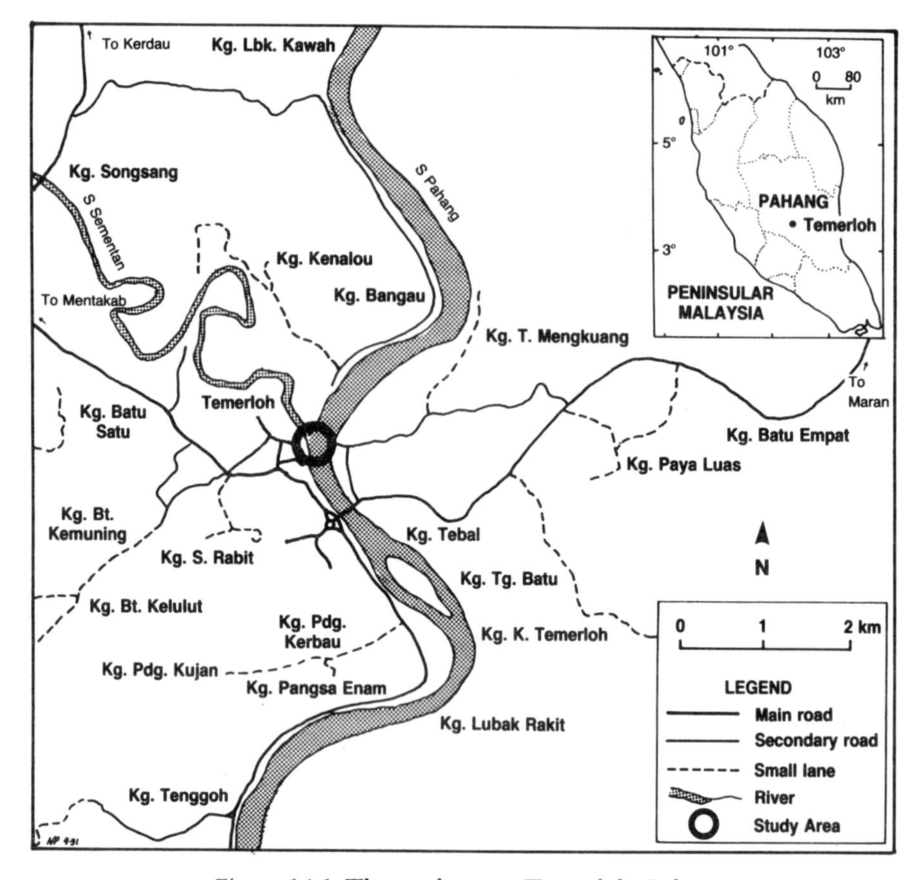

Figure 14.1 The study area, Temerloh, Pahang

THE STUDY

A study was carried out in Temerloh, a town of about 10,000 population, in the state of Pahang (Figure 14.1). The aim was to examine the life of raft-house dwellers, with emphasis on the women who head almost half of the Temerloh raft-households, to see how they fared within and adapted to, their river environment and also to examine their perception of overall quality of life, through the use of a human ecological framework.

A human ecological framework provides a basis for delineating various kinds of quality-of-life indicators, defined as indicator measurements of aspects of human life or environmental conditions which tell us something about the degree of ill-being or well-being of humans and/ or their environments (Bubolz *et al.* 1980: 108). Quality-of-life indicators are functional in assessing the degree at which basic human needs are met, in determining the level of resources or conditions of the environment, and in measuring social conditions of human existence (Land and Spilerman in Bubolz *et al.* 1980).

Based on the general model of the human ecosystem put forward by Bubolz *et al.* (1979), the chapter attempts to understand how the various components of the human ecosystem comprising the human environed unit (the raft-house dwellers) and the environments conceptualised in this study as natural (the river), human constructed (the raft-houses) and human behavioural (socioeconomic activities, etc.) interact with each other in order to achieve a better 'person–environment fit'.

Data were collected in February 1988 by a questionnaire survey of all heads of households of the raft-house community. In total, 32 people were interviewed.

Besides gathering personal, demographic and socioeconomic data on the respondents, information on their satisfaction with their lives in terms of physical and social comforts and security was also collected. The respondents were also asked how they perceived the importance of selected items of life concerns. Perceived satisfaction with life and the importance of certain life concerns were gathered by rudimentary scaling of respondents' subjective feelings and reactions. The placing of ratings on rankings of the degree of satisfaction felt or the importance placed on the life concerns, was made based on the assumption that individuals have a personal hierarchy of needs and values, and can and will place them on a scale from the highest to the lowest.

Items used in the scales were based, selected or adapted from 148 items used by Moller and Schlemmer (1983) with slight modification to suit the local scene. The items used in this study represented life concerns related to human needs, resources of the environment and interaction with the environment.

THE NATURAL ENVIRONMENT: THE PAHANG RIVER

The natural environment is defined as the environment formed by nature with time–space, physical and biological components. These interdependent components constitute the basis for the human life-support system. The natural environment under study is the Pahang River or Sungai Pahang. It is the largest river in Peninsular Malaysia with a length of 435 km. The river exhibits a characteristic flattened profile, being swift and narrow at its headwaters in the mountains and slow, sinuous and broad in the lower reaches, where it finally debouches into the South China Sea.

Most Malaysian rivers and streams are subjected to occasional flooding. The Sungai Pahang, however, floods more or less regularly during

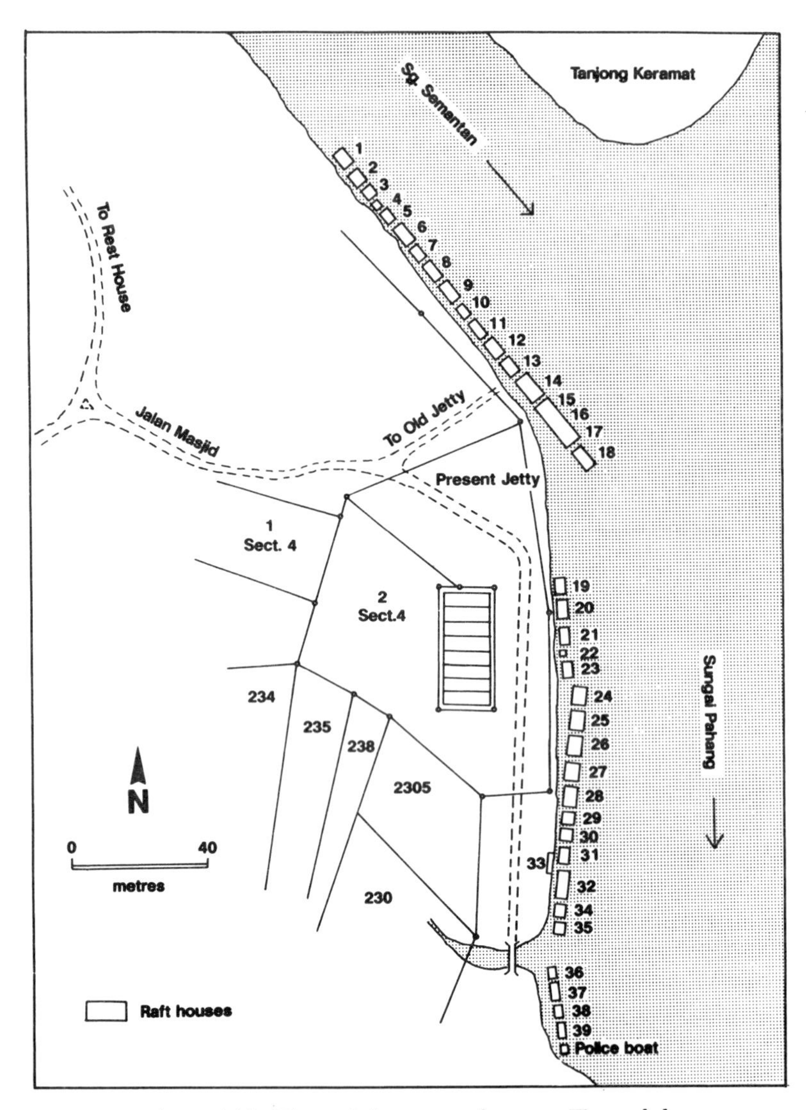

Figure 14.2 The raft-house settlement, Temerloh

the north-east monsoon when very heavy rainfall occurs over a period of about four months (November to February). In the Pahang delta, floods now occur annually, though they were previously only of occasional incidence and the bed of the Sungai Pahang was very much deeper than it is now.

It is at the confluence of the Sungai Pahang and its tributary, Sungai Semantan, that the raft-house settlement is located (Figure 14.2). Flanking the western banks of both rivers, it is sited about 100 km upstream from the main Sungai Pahang. This peripheral location of the raft-houses works to their advantage, that is, reducing their chances of being hit by the direct impact of swift down-stream current flow during flooding.

One of the characteristics of large rivers in Malaysia is the tendency to develop river shores and beaches along their banks due to either erosion, deposition or both. These features are common along the banks of Sungai Pahang. Their presence has encouraged the development of the raft-house settlement at the site, for they provide good landing and anchoring ground thus enabling easy access to and from the banks by foot.

THE HUMAN CONSTRUCTED ENVIRONMENT: THE RAFT-HOUSES

As a human constructed environment, which is defined as an environment altered or created by human beings, a raft-house or *rumah rakit* is a structure that sits on a raft made of either bamboo or logs. The logs, placed at both ends of the house, act as floats, keeping it above water. Rope is then used to hold the raft-house in position and keep it from being swept away by strong currents. Slightly more than half the raft-houses had been in the locality for the past two decades. The oldest was built in 1950, and the most recent in 1986. Chronologically, the history of the raft-house settlement dated to the late 1930s. This is due to some of the oldest inhabitants migrating into the area in 1932. They originally built their houses illegally on government land. When land on which they were squatting was taken away for development by the government, they began to build their houses on water as they did not own any land. The continued existence of this raft-house settlement to the present day is solely because the state government does not consider it to be a problem.

The raft-houses are usually rectangular in shape, with an average size of 11 m × 7.3 m, simply constructed, with wooden floors and walls, and zinc roofs. These houses normally have two doors with the rear facing the river and the front facing the shore. The rear door provides access to a platform-like structure attached to the raft-house. Here the normal daily house chores, such as washing and cooking, are carried out. Some raft-houses have partition-like structures at one side of the rear platform, forming an enclosure, for more private use. On the other hand, the front door is used to go in and out of the house for activities on land. However, there are a few which have a platform constructed at the front door where the household chores are carried out instead. Indoor, most of the raft-houses are sparingly equipped with only a small bed-cum-changing room and such basic necessities as cooking stoves and crockery.

The raft-houses are built side by side at a distance of 1 to 2 metres apart. The raft-house is kept stabilised by an anchor fixed at an angle to one side of the raft. A wooden plank is used to connect the raft-house to the river bank. Where the houses are anchored below a steep river bank, wooden steps are constructed along the slope in order to enable the raft-house dwellers to reach the top of the river bank. Sometimes, steps are cut into the bank's slope.

At places where the river bank is wide enough, some cultivation of short-term food crops such as maize and sweet-potatoes is carried out. Rearing of fowls was also carried out on a small scale, by some, for their own consumption.

Beside human transformations of the physical and the biological environments, other human, social and cultural constructions such as educational, political, and religious systems are included in the human constructed environment. Most facilities such as schools, shops and mosques are easily accessible in the nearby Temerloh Town, and local grocery shops are available a short distance away up on the river bank. An open market for weekly operation is also constructed nearby where most of the dwellers participated.

However, since the raft-house settlement is considered illegal, no basic facilities such as clean water and electricity are provided. As such, all inhabitants resort to oil lamps for lighting, although 15 per cent have generators. The inhabitants also depend on their immediate surroundings, the river water, for both drinking and washing, although there are a few households who normally get a supply of piped water from the grocery shops by paying a certain sum of money for it. Toilet facilities and waste disposal go by natural way down the river.

In terms of ownership, about two-thirds of the raft-houses were owner occupied and the remainder, tenant occupied. On average, the raft-houses fetched about M$50 rental each per month. Among the female headed households, 57 per cent owned their own raft-houses; the other 43 per cent rented them.

THE HUMAN ENVIRONED UNIT: THE RESPONDENTS AND THE RAFT-HOUSE DWELLERS

The questionnaire respondents and the raft-house dwellers constitute the 'human environed unit'. They are a plurality of individuals who have some feelings of unity, share some common resources, goals, values and interests and have some sense of common identity. Altogether, 32 heads of households were interviewed, of whom 44 per cent were females. Together, their households comprised 118 persons of various relationship to the heads. The majority were children, followed by relatives: either siblings, spouses, grandchildren, in-laws and even friends.

Among the respondents, the female heads were found to be older than their male counterparts, 57 per cent being over 50 years of age as compared to 33 per cent of the male household heads (Table 14.1). The female respondents were found to be mostly widows (71.4 per cent) with 14 per cent not married, 7 per cent divorced and another 7 per cent, married but with husbands working and staying elsewhere. The males, on the other hand, were either married (61 per cent) or unmarried (39 per cent).

Table 14.1 Age distribution of the respondents

Age group	*% male*	*% female*
< 20 years	5.6	0.0
20–9	38.9	7.2
30–9	22.2	14.3
40–9	0.0	21.4
50–9	11.1	35.7
60+	22.2	21.4
Total	100.0	100.0

Source: Author's fieldwork 1988

Two-thirds of the female heads had received no education. Among the other third who had been to school, 60 per cent did not complete primary school, 40 per cent attended secondary school but only half completed it. The males, on the other hand, were better educated by at least having some form of formal education with 17 per cent of them having completed their secondary education.

Half of the women heads depended on petty trading for their main livelihood while the other half either depended on wage earning, particularly doing various forms of menial work as factory workers, cooks and cleaning women (36 per cent) or were unemployed. The male heads had a more varied employment structure, with self-employment being more prominent. This included being traders, fishermen, goldsmiths, boatmen (to ferry people from across the river to the western bank) and carpenters. Only 17 per cent took on paid jobs as factory workers, policemen and loggers, 17 per cent were unemployed and another 5 per cent were pensioners.

Almost three-quarters of the female heads of households lived in poverty with incomes below the poverty level of $M300 per month. Only 21 per cent earned $M400 and above per month (Table 14.2). The heads who were unemployed were supported by their children who either stayed at home or were working elsewhere but remitting money back to support the parent. Conversely, only 28 per cent of the male heads earned less than $M300 per month. They were those who were either elderly or who were wage earners especially working in factories. Taken as a whole, 47 per cent of the respondents lived in poverty which compared favourably to the 65 per cent of those in the District of Temerloh identified as poor. This seemingly 'well-off' situation is closely related to the nature of their employment as traders which brings in a better income. However, taken singly, the female-headed households were seen to be worse-off than the general surrounding population. This finding is in accordance with the general findings of the characteristics of the poor, which normally include the aged, widowed and females among them.

Table 14.2 Distribution of income among the heads of the raft households

Income group ($M)	*% male*	*% female*
0–99	5.6	14.3
100–99	0.0	28.6
200–99	22.2	28.6
300–99	55.6	7.1
400–99	5.6	14.3
500+	11.0	7.1
Total	100.0	100.0

Source: Author's fieldwork 1988

Of the other members of the female households, 96 per cent were in the younger age groups of less than 25 years old. Almost three-quarters of them were females. Excluding those who were still in school (24 per cent), 58 per cent of the remaining members were employed as factory workers. This finding supports our earlier contention that urban public housing is unable to keep pace with industrialisation and urbanisation, so leading to squatting; or it may reflect the tradition among immigrants, the Malay especially, to squat or stay with relatives on a semi-permanent basis until such time as an alternative arrangement or accommodation is found; or it could be a combination of both factors. This is especially apparent when further analysis showed that almost half of the household members were actually found to be either relatives or siblings of the heads.

THE HUMAN BEHAVIOURAL ENVIRONMENT

External to the human environed unit is the human behavioural environment which exists because human beings themselves constitute environments for other human beings. It is an environment of socialised human beings and their interrelated behaviour: biophysical, psychological, and social. The human behavioural environment is essential for meeting the physical, biological, social and psychological needs necessary for continuing existence beyond the survival level.

Family, relatives, friends and neighbours all become part and parcel of the daily interaction among the raft-house dwellers. On the other hand, for some of them, especially the household heads, interaction went beyond their immediate surroundings. Their economic activities alone widened their behavioural environment to include clients, customers, employers, workmates, and even the general public. Through this they were able to relate, communicate, learn and find their own self-fulfillment.

COMMUNITY SATISFACTION

Questions regarding satisfaction of physical comforts in terms of settlement type, dwelling, way of life and infrastructure/amenities available were asked. The responses received were somewhat mixed. Half of the female heads were satisfied with their river settlement while the other half showed dissatisfaction. The dissatisfied half were found to be tenant households. Their dissatisfaction is understandable as they only intended to stay temporarily while looking for a better or more suitable area. Almost three-quarters were dissatisfied with their dwelling, quoting size, flood and comfort as the major shortcomings. Accordingly, they preferred to live on land and in bigger houses. They also gave an equally mixed response to the degree of perception of the availability of infrastructure and amenities with lack of water and electricity being quoted often among the dissatisfied. The provision of a stand-pipe in their vicinity would be of great help to them. However, they did not deny the fact that they were conveniently located, being within walking distance of town. Where satisfaction with life was concerned, more than half of them were satisfied with their way of life, while about 15 per cent had mixed feelings regarding it. Comparatively, the male heads were more satisfied than the females where physical comforts were concerned (Table 14.3). The mean scores for the men were found to be generally higher than for the women for all four variables asked. This finding, while not conclusive, suggests two things: that is, either men tended to be more easily satisfied than women, or they could accommodate and adapt more easily to rough surroundings.

Satisfaction with their way of life was closely related to their social and mental comforts. More than 90 per cent of the female heads of households were satisfied with their relationship with their neighbours and the outside community, and with their nutrition. However, more than half of the respondents were dissatisfied with their level of income (Table 14.4). Socially, the female heads were more satisfied than their male counterparts. On the other hand, the men's mean scores for

Table 14.3 Distribution of community satisfaction scores on physical comforts, by sex

	Settlement		*Dwelling*		*Way of life*		*Amenities*	
Scores	*M*	*F*	*M*	*F*	*M*	*F*	*M*	*F*
Very satisfied	16.7	7.1	11.1	0.0	16.7	0.0	5.5	7.1
Satisfied	50.0	42.9	61.1	28.6	61.1	57.2	55.6	35.7
Mixed (don't know)	0.0	0.0	0.0	0.0	5.5	14.3	0.0	14.3
Dissatisfied	27.8	42.9	27.8	64.3	16.7	21.4	38.9	28.6
Very dissatisfied	5.5	7.1	0.0	7.1	0.0	7.1	0.0	14.3
Total	100.0	100.0	100.0	100.0	100.0	100.0	100.0	100.0
Mean score (1–5)	3.4	3.0	3.6	2.5	4.1	3.0	3.3	2.9

Source: Author's fieldwork 1988

Table 14.4 Distribution of mean scores regarding mental and social comforts by sex of heads of households

	Mean scores	
	M	*F*
Relationship with neighbours	3.3	4.3
Relationship with outsiders	3.7	3.9
Income	3.3	3.0
Nutrition	3.9	3.6

Source: Author's fieldwork 1988

income and nutrition were relatively higher than the women's. This is because the female heads were poorer than the male heads, as discussed earlier, hence influencing the standard of their nutritional intake.

Data were obtained regarding respondents' perception of their security, especially in terms of safety from theft. Two-thirds were satisfied with the security services in the area especially during the flood season. The respondents were also asked whether or not they felt safe staying in raft-houses. Fifty-three per cent said they felt quite unsafe due to the possibility of their home sinking, especially when the houses only depended on a rope for stability. They were also particularly concerned about their children's safety especially when the latter's movements and playing space were greatly restricted by the physical environment.

Similar responses were offered by the respondents when they were asked to rate the differences between living in raft-houses during high or low water. Slightly more than half felt that there was no difference in their quality of life during low or high water. Those who felt the difference indicated that low water was preferable because it made washing and other chores easier, although they were cautious that if certain measures were not taken the raft-houses might end up tilted. On the other hand, high or deep water aroused anxiety among them, especially among the more recent dwellers, i.e. the fear of the raft-houses being swept away by the strong currents or even sunk.

The respondents were also asked to assess their satisfaction with their current living conditions. About 60 per cent were satisfied (Table 14.5), with the male heads more satisfied than their female counterparts. The general high degree of satisfaction shown was based on familiarity with living in an aquatic environment (especially among the long-stay residents) and the many positive attributes that the place could offer. The attributes mentioned included: freedom, cheap rent, peaceful, tax free, very convenient by being close to town and other facilities, such as shops, place of work, etc. and with a good and plentiful (river) water supply. These advantages overshadowed the few unattractive characteristics of the settlement.

Being generally satisfied with the ecosystem and the community did not mean that respondents saw no need for improvements. Piped water and electricity were the two most frequently mentioned needs while suggestions such as resettlement and low-cost housing projects, an improved housing environment and house repair were also acknowledged as legitimate needs (Table 14.6).

Table 14.5 Distribution of satisfaction with current living conditions by sex of heads of households

	% M	*% F*
Very satisfied	11.1	7.1
Satisfied	55.6	50.0
Mixed	0.0	0.0
Dissatisfied	33.3	42.9
Very dissatisfied	0.0	0.0
Total	100.0	100.0

Source: Author's fieldwork 1988

Table 14.6 Perception of community needs

Needs	%
Provision of piped water	21.1
Provision of electricity	26.3
Resettlement/low cost housing	5.3
Others	47.3

Source: Author's fieldwork 1988

IMPORTANCE PLACED ON LIFE CONCERNS

Relatively high values were placed on a majority of the items considered as important life concerns, with over 75 per cent of items receiving scores of 4.0 and over, thus indicating a high level of importance.

Several respondents remarked that all of the items were important to some degree.

Attaining a high degree of education, a stable marriage, facilities, children to provide for old age, electricity and sufficient money were the six most important life concerns among the female household heads. On the other hand, the male heads of household rated having their own house, a happy family life, enough money, electricity, stable marriage and physical safety as the most important life concerns to them. Among these life concerns, three were common to both of them although they differed in ranking. These were: having a stable marriage, enough money and having electricity in the house. Owning a car received the lowest mean score and privacy, owning a radio, having peace of mind, having the ability to add to, alter or improve the house, and having piped water in the house also ranked relatively low among the female heads. On the other hand, the male heads rated privacy inside the home as least important, followed by having a radio and the space between houses as relatively less important.

As a whole, the raft-house dwellers did not report a high valuation on life concerns, which might be considered as essential by some people, such as leisure activities and how one spends leisure time. If the respondents are assumed to be valid indicators of valuation, it seems that a leisure ethic was not strong among the raft-house dwellers. They were also less sensitive to privacy at home. The latter is understandable, as structurally, the raft-houses do not allow much room for privacy as most of them have only a single multi-purpose space each.

DISCUSSION

Results of the study can be synthesised to provide a clearer overall perspective of life in raft-houses among the female-headed households in the research area. This will be discussed in relation to: overall quality of life, the environment, and the family and needs and resources.

Overall quality of life

Where overall quality of life is concerned, the raft-house dwellers reported that they were generally quite satisfied with their lives as a whole. Husna Sulaiman and Nurizan Yahya (1986) in their study of 'Provision of housing and the quality of life of the urban poor' in Kuala Lumpur pointed to a similar finding where the residents living in low-cost flats, provided by the City Hall, were generally satisfied with their housing conditions and environment. However, certain sections of the population such as the Malays, the educated and those with expanding family size were less satisfied, often quoting space deficits for sleeping, dining and storage as reasons. In our study, on the other hand, dissatisfaction is associated with the status of ownership and the demographic variables of age and sex. Tenants, the elderly and females were less contented with their living conditions. Living in an aquatic environment is definitely more demanding than living on land, hence only the most resilient will adapt and cope better. The existence of the less resilient only points to the necessity of utilising whatever is available in the absence of other better alternatives.

The readily available water supply is of great convenience to raft-house dwellers. This compares favourably to the experiences of other urban squatter settlements on land, where water has often created major problems, even when stand-pipes are provided. These problems often add considerably to the workload and stress of women in the squatter areas (Azizah Kassim *et al.* 1986). Although, water for washing is of no problem to raft-house dwellers, drinking water needs to be made available through a stand-pipe.

Salience of the environment

On the whole, the raft-house dwellers were quite satisfied with their community, in spite of the lack of several community services. The fact that they are satisfied and happy staying in raft-houses (at least those who are owners) points to the importance of the environment in evaluation of the quality of life. This environment, physical, socio-cultural and human behavioural, provides the basis for satisfaction of important needs and wants, and for interaction with others. The river has provided the basis for cheap housing and convenient basic needs while the family and the general surroundings offer the meaningful interactions important in fulfilling other needs and wants.

Family and resources

The family as a form of the human behavioural environment is found to be of importance to perceived quality of life. Earlier in our study, it was noted that the female heads were mostly widowed, while the males were mostly married. This characteristic has somehow affected the perception of the quality of life among them. Our findings have shown that females had relatively lower scores where perceived quality of life was concerned. Widowed respondents were somewhat less satisfied than those still married. This finding is similar to those of Bubolz *et al.* (1980) who found that those who live alone had the lowest perceived quality of life. Apparently, living alone, without a supporting spouse, children, or others is related to how one perceives one's quality of life.

In addition, the family and hence stable marriage were found to rank very high in importance with our respondents. For many people, putting a high value on family life and great satisfaction with it were closely related, indicating interaction between valuation and satisfaction. The family provides the environment in which physical and other needs are often met. People's perception of their quality of life is a function of how well satisfied they are with what they consider important. What is considered important is a reflection of physical, social and psychological needs (Bubolz *et al.* 1980). Hence, our finding that family life ranked very high in importance supports the above proposition. It can be concluded that family life is one of the most significant components in assessment of perceived quality of life.

CONCLUSION

The use of a human ecological model as a unifying framework for conceptualisation and measurement of quality of life has helped us in understanding how the raft-house dwellers, i.e. the human environed units in the model, perceive their satisfaction with life as a whole as an indicator of quality of life.

Insofar as the raft-house dwellers in particular, and the urban squatters in general, are concerned, it can be said that the existence of comparatively cheap shelter would remain attractive to urban new-comers. This is especially so among the low-income and transient groups. It appears to be a rational choice made in the face of a stiff competitive urban life. Women are no exception to this general rule.

15

DEVELOPMENT AND FACTORY WOMEN

Negative perceptions from a Malaysian source area

Amriah Buang

INTRODUCTION

Manufacturing is indispensable to Malaysia's economic development, not only because of the foreign exchange it earns, but also because of its capacity to generate new employment opportunities. The share of manufacturing employment as a proportion of total employment in Malaysia has risen from 8.7 per cent in 1970 to 15.1 per cent in 1985. A notable characteristic of this has been the massive increase in the participation of women, particularly young, single women from rural areas. This is a phenomenal development in Malaysia since rural women, especially Muslim Malay women, have not previously left their home and family nor their traditional agricultural occupations in the villages, on such a large scale. In 1957, Malay women comprised only 6.6 per cent of the manufacturing sector's labour force while Malay men made up 66.6 per cent; by 1976 women's share stood at 18.8 per cent and men's participation had dropped to 47.9 per cent. By 1979, the number of Malay women in the female labour force exceeded that of Chinese women who had previously been predominant (Jamilah 1984: 78–9).

The majority of Malay women factory workers are rural-urban migrants or originate from a rural background. Most of them belong to large families with low incomes. They migrate to reduce economic dependency on their families and with the hope of being able to remit money home. With prospects of better monetary incomes and material well-being, many women have gone to work in the alien occupational and social environment of multinational companies, creating a structural change in the employment pattern of Malaysia (Ackerman 1980). This change was encouraged by the implementation of the ethnic quota imposed by the New Economic Policy (NEP) after 1970.

NEGATIVE PERCEPTIONS AND MALAY FACTORY WOMEN

Negative perceptions regarding the 'socio-moral' consequences of the employment of Malay women in factories is not a new issue. In 1978, newspapers reported illegitimate pregnancies among factory women, incidents of abandoned new-born babies in industrial communities and the involvement of female factory workers in social activities which conflicted with traditional Muslim Malay cultural norms and values (*Utusan Malaysia* 10 and 29 October, 1978). Since then, the negative effect of factory employment on Muslim Malay women has been a cause of concern and debate in Malay society.

In 1979, Jamilah conducted a survey in rural communities of attitudes to various aspects of the migration of Malay women to factory employment. In this survey, parents with daughters currently employed as factory workers in the cities were interviewed about the circumstances leading to their daughters' employment and their knowledge and opinion of the social position and reputation of factory women in society. Married couples with daughters working in the factory while retaining village residence were also interviewed. Again, they were questioned as to their opinions of female outmigration to the factories and the reputation of factory girls.

Jamilah's results tabulated from 2,000 village respondents, representing 60 villages throughout Peninsular Malaysia, may be summarised as follows:

> many parents of factory girls were worried about the morality of their daughters. They were also beginning to feel ashamed that these girls were employed in an occupation which was rapidly acquiring a low moral and social status in Malaysian society.
> (1984: 240)

Malays who were concerned about the moral failings of women factory workers were not restricted to parents, but included prominent women politicians who urged rural parents not to allow their daughters to migrate to urban-based factories, for fear that they might be corrupted by the permissive culture of multinational firms.

Members of Malaysian urban communities have similar negative perceptions. Jamilah concluded the findings of her urban community survey in 1979 and 1981 and her longitudinal study of the period 1977–83 as follows:

> The findings . . . confirm that generally migrant Malay factory girls are regarded as of doubtful morality. They have low moral status in the urban–industrial community. The . . . urban residents indicated clearly that factory girls, living on their own, lead undesirable

life-styles; they are not willing to let their children mix with these girls . . . they believe many factory girls are women who can 'easily be led astray by bad elements'. They would think twice in selecting factory girls as potential wives and even if they decided to accept a 'decent' factory girl as a bride, they would insist that she stops working in the factories after marriage. They do not wish their wives to be associated with an occupation which has a low social and moral status in Malaysian society. Generally, the older residents in the urban–industrial communities have a more hostile and suspicious attitude toward the migrant factory girls

(Jamilah 1984: 253–4)

NEGATIVE PERCEPTIONS FROM PERMATANG PAUH

For the purpose of examining the perceptions of a local community regarding the employment of women in factories, a study was conducted in Permatang Pauh in November 1982. Permatang Pauh is an area lying adjacent to the industrial zones of Seberang Prai, the peninsular part of the state of Penang in the north-western district of Peninsular Malaysia. Permatang Pauh (Figure 15.1) consists of 50 villages connected by a road network to three major industrial zones in the Penang–Seberand Prai state: the Mak Mandin, the Bagan Serai and the Prai. The distance between the villages and these industrial complexes varies from 6.4 to 24.1 kilometres.

The study included interviews with 1,468 adults (mainly heads of families) from 45 villages. All the respondents were dependent on rice farming as their main source of livelihood, although at the time of the survey a large number of interviewees had ceased farming their rice fields due to repeated seasons of bad harvests. Of the total number of respondents, only 258 (17.6 per cent) admitted to having at least one female family member working in a factory on a permanent basis. More than three-quarters of the total sample declared that none of their female family members were involved, while a small percentage (5.5 per cent) declined to answer. Of those families with female members involved in factory employment, the majority (174 households) had only one female household member in factory employment.

The low degree of involvement of the respondents' own family members in factory employment indicates that they espoused a dual position with respect to their perception of women factory workers in the nearby industrial areas. First, as an involved party evaluating the changes resulting from the employment of their relatives in the factories, and second as a detached observer, evaluating the changes resulting from the employment of their neighbours and of migrant female workers who sought accommodation in the villages fringing the industrial area.

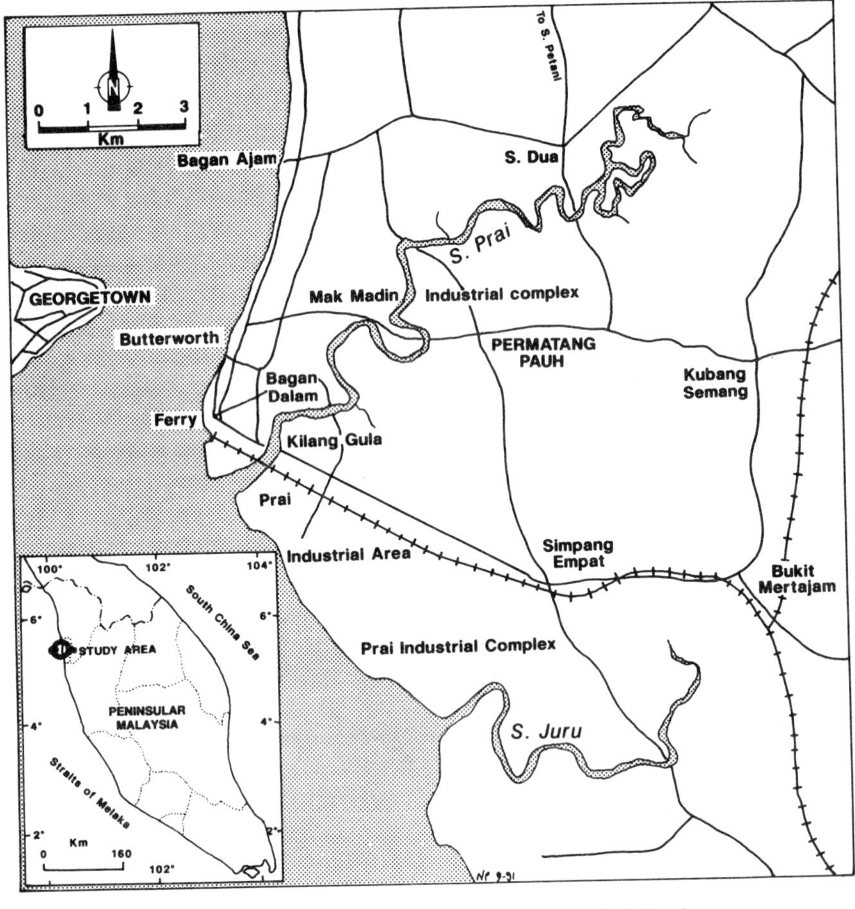

Figure 15.1 The study area, peninsular Malaysia

Interviews with household heads indicate that the source communities were unhappy with what were perceived as deteriorating moral standards among factory women. These pertained to the women's manner of dressing, which was judged 'indecent', their social mixing, which was rated 'very liberal' and 'permissive' and their decreasing interest in local affairs. Together these factors account for 40.7 per cent of the total responses (Table 15.1).

In contrast, positive responses such as improvement in household living standards, increased knowledge among the factory women as a result of greater social exposure and a higher level of self-sufficiency, were less common answers. Such comments constitute 11.9 per cent of total responses indicating a weak support for female social and economic independence. Negative perceptions were voiced three and a half times more often than positive comments.

Table 15.1 Perception of personal changes that occurred when women became factory employees

Type of change	*Number of respondents*	%
Liberal socialising	322	21.9
Indecent dressing	143	9.7
Decrease in moral standards	126	8.6
Improved standard of living	100	6.8
Increased knowledge	66	4.5
No noticeable changes	55	3.8
Self-supporting	9	0.6
Less interest in local affairs	5	0.3
Devaluation of domestic role	4	0.2
'Difficult to say'	512	35.0
No comment	126	8.6
Total	1,468	100.0

Source: Author's fieldwork 1982

The proportion of negative perceptions may be higher than the statistics indicate if we take into consideration the 512 respondents (35 per cent) who were classified in the 'difficult to say' category (Table 15.1). This group stated their reluctance to incur the anger of factory women and their families as the reason for their difficulty in describing the changes in specific negative terms. The source community were also of the opinion that there were alternative, more suitable occupations for women, such as school teaching, clerical work, nursing, handicrafts, farming and managing a business (Table 15.2). For 42.3 per cent of the respondents who were uncertain whether there were suitable alternative occupations for women, home making was their preference. Reasons for a preference among alternative non-factory jobs (Table 15.3) centred around economic remuneration (49.7 per cent). Social, personal security and morality reasons constituted another 49.1 per cent of total responses.

Table 15.2 Occupations considered suitable for women

Occupation	*Responses* *Number*	%
Government jobs (school teachers, clerks, nurses)	294	20.0
Self-employment (handicrafts)	163	11.1
Agriculture (other than rubber tapping)	136	9.3
Business	91	6.2
Home maker	41	2.8
Rubber tapping	10	0.7
Housemaids	6	0.4
Uncertain	727	49.5
Total	1,468	100.0

Source: Author's fieldwork 1982

Table 15.3 Perception of the suitability of alternative occupations

Perception	*Responses* *Number*	*%*
Pension benefits	191	25.8
Higher income	177	23.9
Greater safety	102	13.8
Less morally detrimental	96	13.0
More fitting with home role	91	12.3
Better working conditions	75	10.1
Match with school qualification	3	0.4
'Uncertain'	6	1.0
Total	741	100.0[a]

Source: Author's fieldwork 1982
Note: [a] Some totals are rounded

Since these broad groupings are approximately equal we can assume that judgement does not appear to be wholly influenced by a particular stereotype.

The proportion of responses which supported the idea of encouraging Malay Muslim women in general to embark on a life-long career in the factory sector is numerically larger than the converse (37.3 per cent compared to 28.9 per cent: Table 15.4). When the same question was posed about the respondents' own female family members, the percentage of positive responses fell to 10.9 per cent (Table 15.5), a difference of 70.8 per cent. The question therefore arises: why is it that 71 per cent of those who supported the idea of long-term employment of women in the factory sector change their opinion when the women workers in question are members of their family? This situation allows us to summarise the sample case study as follows: (1) the source community supports, in general, the employment of women in factories; (2) nevertheless, it does not favour factory employment on a long-term basis; (3) factory employment, on a permanent basis is definitely out of the question if it directly involves a member of the respondents' own family. Thus, it seems clear that the phenomenon of female factory employment is something that is tolerated rather than welcomed in the source area. The interplay of the undesirable 'side effects' of factory employment and the immediate economic benefits from it are responsible for producing this rather ambivalent attitude.

A comparison of the data in Table 15.3 with that in Table 15.1 suggests that while short-term considerations prompted the support of the source community for factory employment of women, long-term perceptions prompted the opposite. The short-term considerations were mainly economic while those of the long term were the combination of economic (remuneration and better working conditions) and sociomoral factors, with the latter approximating the former (Table 15.3). Thus,

Table 15.4 Perception of whether Muslim Malay girls should be encouraged to work in factories

Perception	*Respondents* *Number*	%
Should be encouraged	548	37.3
'Difficult to say'	433	29.5
Should not be encouraged	424	28.9
Depends on circumstances	37	2.5
No comment	26	1.8
Total	1,468	100.0

Source: Author's fieldwork 1982

Table 15.5 Perception of whether respondents' female family members should be encouraged to work in factories

Perception	*Respondents* *Number*	%
Depends on circumstances	268	18.3
'Difficult to say'	211	14.4
Should be encouraged	160	10.9
Should not be encouraged	133	9.0
No comment	696	47.4
Total	1,468	100.0

Source: Author's fieldwork 1982

even though social and moral considerations did bear rather heavily in the long run (Table 15.1), immediate economic pressures forced the toleration of factory work as a temporary solution.

The case study allows us to make a number of points regarding the perception of factory employment for women among the people of Permatang Pauh. The community is aware of the economic benefits of factory employment; the community is also aware of negative aspects, namely the deterioration of moral conduct and social responsibilities. Nevertheless, the community concedes the short-term benefit of factory work as a practical solution to the households' immediate economic needs. In the long term, however, respondents do not see the work as a solution to economic needs, especially in light of the moral questions that emerged. Finally, the community would reject both short- and long-term factory employment if there were alternative solutions to its immediate economic problems.

JUSTIFICATION FOR NEGATIVE PERCEPTIONS?

Several researchers have expressed resentment and doubts over the negative perception of factory women by various segments of society (Jamilah 1984; Fatimah 1985). They attempt to identify factors that underlie the 'unjust' attitude of Malaysian society towards factory women. Jamilah deplores the press publicity received by illegitimate pregnancies and the moral conduct of factory women. It is based, she claims, on hearsay and gossip and on reports from a minority of factory women and members of the urban–industrial communities, rather than on established facts. Jamilah believes that there is inherent suspicion among conservative elements in Malay society regarding single women residing outside the supervision of their elders.

Jamilah rationalised the political, economic and racial reasons that underpin the urban community's resentment of the presence of factory women, thereby making factory women 'societal scapegoats'. The establishment of free-trade zones in Penang's Bayan Lepas was at the expense of several vegetable farmers who lost their land, and of squatters and residents who were forced to evacuate their homes. As a result, the immigration of the women workers was identifiable with the developers' interests and was thus viewed with resentment by local residents. In addition, the influx of migrants prompted a sudden rise in the prices of foodstuffs and rental accommodation. Irate residents began to express their anger at these changes in the form of hostility and unfriendliness towards the migrants. Similarly, the local Chinese residents of Sungai Way Free Trade Zone in Selangor suspected that there was ethnic bias in the government's decision to develop the area into a free-trade zone, establishing factories which were allowed to recruit Malay workers in order to conform with the New Economic Policy. The suspicion that this was an indirect method of weakening the Chinese hold in the area made them less hospitable to the Malay factory migrants. Nevertheless, regardless of whether the local residents were Malays or non-Malays, they were not accustomed to having young, single women workers as their neighbours. The women were regarded as a threat to the established social order (Jamilah 1984).

Fatimah reasons that the moral stigmatisation of factory women is related to male chauvinism. Women in Malaysia are regarded as dependent on and inferior to men. The man is always the head of the household, the breadwinner and the protector. The factory women had reversed these roles by contributing to family income and gaining independence. Therefore, some male members of society viewed the participation of women in industry as emasculation and expressed their 'anger' towards females engaged in 'men's jobs' by accusing them of immoral behaviour. Fatimah suggests that similar immoral behaviour among clerks, typists, and other occupational groups is ignored because in these instances, the workers are employed in 'women's work'. She feels that some men cannot accept the attitudes of employers who willingly employ more women than men in the manufacturing sector. Thus, women (and their male employers) constitute a threat to male egos and men retaliate with vitriol and contempt (Fatimah 1985).

Apart from men's prejudices, Fatimah also indicates that the stigmatisation of women factory workers coincides with the image of the urban–industrial area. Many of the factory women moved from rural to squatter settlement areas. The latter are regarded as crime-ridden and unhealthy (Kraal 1979; Zainah 1978).

Both Jamilah's and Fatimah's arguments seem to indicate that the low moral status accorded by Malaysian society to Malay factory women is not justified. The question which needs to be addressed, especially in light of the surveyed community's negative perceptions, is whether there is really any justification for society's attitude. This is addressed by empirically examining the reality of factory women's working lives and by evaluating these findings in terms of the normative value system upheld and subscribed to by both the factory women and general Malay society.

WOMEN AND FACTORY WORK

Jamilah (1984; Grossman 1979; Khoo and Khoo 1978; Lim 1978) reported observations of the American multinational sociocultural system and its impact on Malaysian culture. American companies have a Western-based sociocultural system. Many companies organised beauty contests where female workers vied for awards and prizes which clearly encouraged them to become 'sex symbols' and adopt liberal social mores. For example, production operators who won beauty contests were usually promoted to the role of the company's receptionist or social escort for company guests and overseas visitors. The prizes offered in the contests included free courses in grooming, overseas trips and often overnight hotel reservations for two. The women workers were encouraged to invite their boyfriends to annual beauty contests and balls. At these functions, alcoholic drinks were freely available. Female workers were also encouraged to join outdoor sports and athletic activities where men and women participated jointly.

As employees, the Malay workers had to adapt to these cultural demands in full knowledge that they were transgressing Islamic injunctions. It should be pointed out that although not all multinational companies forced their workers to participate in such activities, most workers were urged to do so. Inevitably, workers had to adapt if they wanted favourable recognition from superiors. In addition, many factory women, being young and single, were eager to experiment with new ideas and thus were easily influenced by what they perceived as the normal practices of modern (Western) people and the modern (Western) way of life. Indeed, many migrant girls perceived modern urban life as revolving around dating, dancing and alcohol (Jamilah 1984). Given the imposed Westernised cultural atmosphere and the competitiveness among the female workers, it was very common to observe attempts by the women workers to dress according to the latest fashion and behave like Americans. Several wore 'Farah Fawcett hairstyles, tight blue jeans and thick facial make-up. When the mini-skirt was in vogue, they were hitching up their work uniforms to a mini-skirt level' (Jamilah 1984: 239).

The tendency for several Malay women workers to reside without the supervision of parents and relatives encourages visits from men. Free from the strict supervision of their families, some of the women entertained men at their homes. In some cases, parties were organised at which modern, Western-type dancing took place. In the period 1977–8, a number of women were caught with their boyfriends in public places and charged in the Islamic law courts with indecent promiscuous behaviour ranging from *khalwat* (close intimate proximity between unmarried couples) to *zinah* (illicit sex). Some factory women were daring enough to co-habit with men. In some cases, this and other relations with men resulted in unwanted pregnancies and illegitimate children (Jamilah 1984).

Conducting research in a Japanese-owned company, Fatimah reports similar findings. In addition to the big annual carnival show consisting of drama, songs, dances and beauty contests, there were smaller parties held regularly throughout the year. These came complete with rock bands, and the participants had ample opportunity to demonstrate their knowledge of the latest dances at such times. Attendance at the parties necessitated some knowledge of dancing or at least a willingness to learn (Fatimah 1985). There was a conscious development of a middle-class feminine ideal. The young women were encouraged to see themselves as 'sex objects' through the promotion of beauty or 'sweetheart' contests, fashion shows and beauty culture classes. Related to this was the development of a consumer culture aimed at the same feminine ideal. For example, every week make-up representatives from Yardley and Max Factor visited the firm. Workers became the victims of hire-purchase 'con-men' who could be seen at their door each pay-day collecting the 5- or 10-*ringgits* instalment for dresses and cosmetics purchased on credit (Fatimah 1985).

Research conducted by Jamilah and Fatimah is not hearsay, gossip or conjecture, but provides facts about the practices, norms and values built into the Malay factory women's working lives which clearly contradict their religious injunctions. The normative value system upheld and subscribed to by Malay Muslim society in general depicts the Quranic view of the chastity of women:

> Say to the believing men that they should lower their gaze and guard their modesty, that will make for greater purity for them . . .

> And say to the believing women that they should lower their gaze and guard their modesty; that they should not display their beauty and ornaments except what must ordinarily appear thereof; they should draw their veils over their bosoms and not display their beauty except to their husbands, their fathers, their husband's father, their sons, their brothers, and their brother's sons or their sister's sons, or their women, or the slaves whom their right hands possess or male servants free from physical needs, or small children who have no sense of sex; and that they should not strike their feet in order to draw attention to their hidden ornaments. O ye believers! Turn you all together towards God, that you may attain bliss.
>
> (24: 30–1).

> For men and women who guard their chastity . . . for them has God prepared forgiveness and great reward.
>
> (33: 35)

> Women impure are for men impure, and men impure are for women impure, and women of purity are for men of purity; and men of purity are for women of purity; these are not affected by what people say; for them there is forgiveness and provision honourable. (24: 26)

> The woman and the man guilty of adultery or fornication flog each of them with a hundred stripes: Let not compassion move you in their case in the matter prescribed by God if you believe in God and the Last Day. (24: 2)

> The believers, men and women, are protectors, one of another; they enjoin what is right and forbid what is evil.
>
> (9: 71)

These Quranic verses clearly prescribe that Muslim women must dress decently, safeguard their personal modesty, avoid permissive and promiscuous socialising and must not be treated by themselves or others as sex symbols. Muslim Malay society accepts these Islamic dictums and subscribes to them.

In light of Muslim Malay society's beliefs, the Permatang Pauh source community's perceptions were quite reasonable in their moral expectation of the Malay women factory workers. It is not the fact that the women went out to work in modern factories that people resent, rather it is what they perceive to be an outright transgression of what they have cherished as good and decent that they are unwilling to tolerate. On principle they cannot accept any form of religious transgression committed by any occupational group of Muslim women. They will not, for instance, approve of similar acts of transgression by Muslim female teachers, clerks or nurses. However, the cases of factory women's misconduct appear to have received wider publicity, evoking public reaction, due primarily to the number of women working in factories and the sociomoral systems of foreign-owned factories.

The proportion of Malay women factory workers rose from 19.1 per cent in 1975 to 26.2 per cent in 1979 (Labour Force Survey Report 1979). Furthermore, the nature of factory production results in a spatial concentration of the workforce. These factors make the presence of women workers highly visible and conspicuous and so the impact is readily felt in specific neighbourhoods as well as society at large. The problem is compounded by the fact that the foreign-owned multinational companies which employ the bulk of Muslim Malay women, do indeed articulate values and rituals that conflict directly with Islamic ideals. The teaching profession in Malaysia does not preach indecency, nor make it a point to groom beauty queens and 'sex symbols' or to hold parties. The Malay clerks and typists, the majority of whom are found in government offices, are subjected to the public service ethical codes in which decency in dressing and social manners are strictly required.

Society's attitude to the Malay factory women has led several multinational companies to amend their recreational activities in order to avoid public controversy. Beauty pageants, annual balls, Western-type dancing and the consumption of alcohol are now a rarity and male and female sporting activities have become segregated. However, very little progress has been achieved with respect to the moral de-stigmatisation of Malay factory women. The Muslim Malay society is not convinced that this new morality of the multinational companies is a genuine matter of faith and conviction and not merely artificial devices to pacify the women's parents while attracting their daughters to work in the factories. Hence, the continued suspicion and scepticism of Malay society about the impact of factory employment on Muslim women.

In this regard, it is interesting to note the interpretive ability of society. The source community may not understand the broad structural factors, both international and local, underlying the phenomenon of factory women that Jamilah sought to explicate. They may not know that the interplay of forces encouraging the establishment of export-oriented industrialisation with the implementation of the New Economic Order has resulted in a sudden expansion of employment opportunities for Malay female workers. Although the community could sense that worthwhile job opportunities were becoming scarce in the rural area, they knew very little of the exploitation of the industrial proletariat by their capitalist multinational employers that Fatimah (1985) was eager to highlight, nor of the capitalist tendencies to 'intensify', 'decompose' and 'recompose' women's subordination at the industrial workplace that

Elson and Pearson (1981) conceptualised. However, the community definitely knew something of the 'feminine' false consciousness that was being reinforced by factory employment. They knew that the factories encapsulated their female workers into alien social activities and subcultures.

Judging from the answers given by respondents in Permatang Pauh to questions regarding the reasons for their low opinion of the factory working environment, it is clear that the source community was aware of the moral dangers and risks facing the Muslim factory women. They (70.3 per cent) perceived that young women were handicapped in coping with the urban–industrial alien and threatening way of life. Sixty-nine per cent of the respondents believed that the women did not know how to handle interactions with men and might easily fall prey to prostitution racketeers and ill-intentioned men. At an impressionable age, and in an eagerness to experiment with what they were taught to think of as the normal way of modern and urban life, 74.5 per cent of the respondents felt that the women could easily be influenced by bad elements in the urban–industrial community. All the respondents believe in the sanctity of marriage and want their daughters or female relatives eventually to marry. However, 92 per cent of them did not want a situation where the women compromise their moral integrity and transgress their religious injunctions in their search for a husband. In this vein, 79 per cent of the respondents strongly opposed the idea that unwanted pregnancies could be avoided by teaching the factory women about contraception. To them, *zinah* (illicit sex) is still *zinah* whether it results in pregnancy or not.

The source community also had definite ideas as to who should be held responsible for the moral corruption and stigmatisation of the Malay factory women. The majority of them (79 per cent) blamed the multinational value system and socialization structure, although 65.5 per cent did not relieve the women of personal responsibility. Every Muslim must make efforts to protect themselves from transgressing the religious injunctions, especially when related to grave sins such as *zinah* and moral decadence. Perhaps such a stance underlies the observation made by Jamilah (1984) and Fatimah (1985) that many factory girls have become reluctant to participate in activities organised by the factory management.

PROSPECTS AND CONCLUSIONS

The moral stigmatisation of Malay female factory workers is likely to continue as long as Malay Muslim society is not convinced that the factories are a mainspring of that which is Islamically decent and good. To achieve this the foreign-owned companies must be willing to perform more than just cosmetic adjustments to the sociocultural activities and values that they institute. For instance, instead of abolishing beauty queen contests, annual balls and parties, they could improve their image by sponsoring religious classes, talks and courses, appropriate educational programmes and other activities geared to make Muslim women aware of their multifaceted role in society. Similarly, instead of merely ceasing to persuade and urge Muslim women workers to participate in Western-type cultural and social activities, the companies would be more convincing to the local Muslim society if they embarked on policies and activities that would not only reflect their respect for the host society's Islamic values but also their seriousness in safeguarding and promoting them. These may range from designing Islamically decent uniforms and providing prayer rooms in the factories, to offering facilities for the organisation of funds for welfare purposes which are in line with Islamic principles.

The prospect of adoption of such suggestions is not very great. This is not because organising beauty contests, annual balls and other activities is less expensive, but because the companies have the choice of not needing to do so. Relocating their factories in other countries which provide equally cheap female labour but which are less culturally and religiously demanding may be advantageous. Alternatively, they may be allowed to ignore the expectations of Malaysia's Muslim society as the nation is in urgent need of strengthening its economy through industrialisation.

Thus, as long as the presence of the multinational companies is wanted in Malaysia, and as long as Malay Muslims uphold their religious teachings, there will be no complete happiness for the Malay women factory workers. Those who resist the encroachment of the alien culture in the factories will continue to face problems in their working relationship with their peers and management, and those who embrace the new culture will continue to suffer from the moral stigmatisation of their society.

16

WOMEN AND AGRICULTURE IN WESTERN SAMOA

Peggy Fairbairn-Dunlop

> Everytime the statistics men come, I list myself as 'Planter'. When the role comes out, there I am 'Housewife'! I tried to see where the change was made. I never could.
>
> (Woman Planter 1988)

> Social conventions affect the enumeration of women in general. A woman's decision about enumerating herself as part of the labour force is very difficult, and is usually in accordance with the norms of society. If the culture is one where women do not participate in labour force activities then women could tend to report themselves as having done no productive work. This traditionally has been the case in Western Samoa.
>
> (Salale 1976: 92)

INTRODUCTION

These comments illustrate the attitudes concerning women and women's place in Samoa, attitudes that are also evident in other South Pacific countries, such as Tonga. The first comment illustrates that women's position today is one of contributions ignored, as men place women in a role that concurs with the *faaSamoa* ideal, an ideal based on the principles of the precedence of dignity and achievement of the Samoan *aiga* (family) over individual achievements, and of the *feagaiga* ideal (sacred covenant) whereby brothers are responsible for the economic, social and political well-being of their sisters, so that their sisters do not have to labour.

The second comment, presents a picture of rights deferred by women themselves, for women are also bound by the traditional ideal of family status. Of women being willing to be seen in a 'supportive' non-productive role, for to indicate otherwise would be to acknowledge that their *aiga* cannot or is not fulfilling its traditional function. This would be both personally and socially unacceptable.

WOMEN IN SAMOA

The women of Samoa are caught in a bind between the old and the new, between the traditional ideal that women do not do agricultural work (born in a time of *aiga* and village economic self-sufficiency), and the actual situation emerging out of new aspirations forged in an increasingly monetised, individualistic and materialistic society, a society where the traditional base is losing its meaning but where many of the forms are vigorously defended. This dilemma is all the greater because Samoa has established a reputation for extraordinary cultural conservatism (Holmes 1974).

Present policies and opinions regarding women's participation in development in Samoa, are marked by traditional perceptions of women's role in society. Given the absence of statistics to indicate change, the lack of women's input into government due to *matai* (chiefly) suffrage and no recognised channel for communication of women's opinions, development planning continues to be built on these assumptions.

It would be difficult to pinpoint the true source of these perceptions of women's place, for they represent a blend of indigenous and European beliefs. Traditionally, women and women's groups had enjoyed considerable autonomy of action and influence, related to traditional avenues of power and expected roles. Colonial administrator, missionary and trader did not see or acknowledge this autonomy and independence. Women and women's groups were excluded from participation and decision making in the formal sphere in the emerging forms of government. Their opinions and input were confined to certain types of activity, for example, health and family welfare:

> Agricultural work was seen by many colonial officers and administrators to be inappropriate, even degrading for women and it was believed that their status would be enhanced if their roles were exclusively that of providing domestic services for their children and husbands.
>
> (Schoeffel 1986: 42)

It is paradoxical that, at the same time as colonial authorities were teaching that women should not be doing agricultural tasks, the social and economic changes taking place, such as the introduction of modern consumer goods, schools and cash cropping, created new aspirations and needs. These made it almost inevitable that women's contribution to the household production unit would both increase and diversify, in order to achieve and sustain the desired lifestyles.

Land and labour have always been Samoa's chief resources. There are no mineral deposits, and distance from major trade routes inhibits industrial development as well as limiting market opportunities. Increasing agricultural production both for national food sufficiency and cash-earning potential, has been the major aim of development policies, and the guaranteed markets of the colonial period ensured that production did increase.

More recently, agricultural production has declined, despite heavy government investment; a trend noted for other South Pacific nations also. Export earnings from the agricultural sector averaged between 85–96 per cent of the nation's total earnings between 1977 and 1980, but 1981 figures were 78 per cent with the declining trend 'thought to be continuing'. On the other hand, imports have quadrupled between 1975 and 1985, and are continuing to increase. The most serious aspect of the increase in imports has been the growing dependency on imported food. In 1983, food was the major import, accounting for 22 per cent of total imports. This trend, has been associated with a decline in nutritional and health standards (Thomas 1987; Sio 1987).

There are many reasons for this lack of agricultural growth, but a major one is that development planning has failed to take into account the realities of the smallholder sector, in particular the composition of factors affecting the available labour force (Fairbairn 1985).

Agricultural planning today proceeds on the assumption that the extended family is still the main unit of production in the villages, and more specifically that it is the males in the family unit who have major responsibility for agricultural production. Old assumptions about work roles prevail. These are based on a division that attributes work that is light, clean and focused on the central village to women, while heavy, dirty tasks and those associated with the bush or areas peripheral to the central village area are more clearly men's work.

Statistics collected according to imported Western models, based on concepts of work and economic activity derived from an advanced market economy, reinforce the notion that women's contribution to the economy of the household and indeed of the nation is insignificant. Current statistics list 78 per cent of males as being 'economically active' while only 14.6 per cent of females fall into this category. Commenting on these figures, Thomas suggests that they point clearly to the problems of classifying women's work, as the small proportion of women reported as being economically active bears no resemblance to rural realities. In rural areas young women help with agricultural work and with community-based agricultural and social projects, and older women spend up to half their daylight hours weaving mats which are an important source of cash as well as a major item in the traditional economy (Thomas 1986).

Yet these are the figures on which development planning is based. Consequently, agricultural information and support systems continue to be directed towards men. Neither women, nor women's committees have been given a place in rural development planning. All social and economic planning at village level is channelled by government through the *pulenuu* (village mayor) to the village council of chiefs, (*fono*) which is predominantly male:

> No corresponding recognition is offered to the President of the Women's Committee, nor do direct channels of communication exist between government planners and the committees, except, in theory, through the *pulenuu*.
>
> (Schoeffel 1986: 9).

Therefore, plans and projects related to what is seen to be women's sphere of influence, such as sanitation and water-supply projects, district hospital development, agricultural projects and marketing schemes, are developed without consulting or sometimes even informing the women. Yet, 'rural women are expected to co-operate enthusiastically with whatever role, if any, has been devised for them' (Schoeffel 1986: 10).

Accurate data concerning women's contribution to the economy of the rural household and the factors which affect this must be obtained. For Samoa, the main question is, what is the composition of the labour force in the villages, and what are the factors that determine this?

After the introduction of the monetised economy, migration is probably the main factor to which the rural household has been forced to adapt. The increased efficiency of transport and communication systems in the post-independence period, has brought about an acceleration of the 'rural exodus' to town, and overseas. Samoa registered a declining population growth rate in 1981 (0.6 per cent in 1981 compared with 2.2 per cent in 1976) and emigration reached a high level of 1.8 per cent in 1986. This migration is age-specific (15–29 years) and dominated by males. Urban job commuting, either daily or weekly, is another factor drawing vital labour out of the rural areas, and has been a feature of Samoan life for at least three decades (Brookfield and Ward 1988). Again, males predominate in this process.

A common saying in Samoa today, is that the rural villages are 'the places of women and children'. And this is so. For example, in the study to be discussed, five women 'heads of households' were noted, and other information suggested that many 'younger' women were married to much older husbands. The village population was atypical in other ways also. Fifty per cent of the population of 900 was under 14 years of age, and roughly 24 per cent was in the 15–49 age group.

Many of the school-aged children (5–19 years old) were in school for most of the 'working day'. There was thus a reduced labour force whose task it was to increase food production, to meet the daily needs of the family and, if possible, produce a marketable surplus, as well as fulfilling the 'reproductive' needs of the family.

It has been argued that the remittances associated with migration provide the major income for rural families, so that they do not have to work hard to increase agricultural production. Remittances are large: official remittances for 1983 for example, were estimated at $31.5 million, while unofficial inflows were estimated to be more than double this figure. Remittances, however, do not replace the need to farm. Remittances are not regular, nor are they evenly spread among families in the villages. Villages are fast becoming scenes of increasing inequities in wealth and lifestyles and, with the growth of economic individualisation, many of the traditional reciprocal support systems are losing their meaning (Schoeffel 1984).

For most families, agricultural production is still the major source of food and income, and for many, the only source of income. In addition, increased agricultural production must be achieved with a decreasing workforce. In-depth studies of the operation of rural households are necessary to provide data about what is being done, who is doing it and why. These data will help in devising future development plans and educational support systems, as well as bringing recognition of women's contribution to the economy of the household, village, and nation. The case study discussed below is a first step in that direction.

CASE STUDY: WOMEN'S AGRICULTURAL WORK IN THE SMALLHOLDER SECTOR

This study, undertaken in 1986, focused on women's contribution to the household agricultural production unit, and women's perceptions of their contribution. The purpose of the study was to document the nature and extent of women's work and the factors affecting this contribution.

Over a 1-week period, women respondents from each of 54 households comprising a village were visited and interviewed daily by one of ten recorders from the School of Agriculture. Because only one village was studied, for a limited time and at a particular time of year with no allowance for seasonal variations in work patterns, only limited generalisations can be made. However, the study did generate further questions which would benefit from further study on such issues as land rights, distribution of income and how household decision making occurs.

The village

The sample village was a land-rich village, with land stretching in a wedge shape back from the very wet coastal area, up to high mountains where temperatures are cool. As with all customary land in Samoa, land is held by the *aiga* (extended family group) under the control and trusteeship of the family *matai*. As female heirs to the *matai* title, sisters have rights equal to male heirs concerning access to and use of family lands. But, traditionally, sisters have rarely claimed land for themselves, since its tilling and cultivation is not regarded as the responsibility of the daughters of the *matai*. If a sister marries, and decides to settle with her own family, she and her husband automatically have access to her family land (Aiono 1986).

The village had the appearance of a thriving, reasonably wealthy community. There were few European houses and Samoan *fales* (houses) predominated. Church structures were large and impressive and a two-storey school block had recently been opened, funded by a combination of village and overseas donations. Women ran the four small shops in the village, and a post office. Two women took in sewing and two provided sugar cane and pancakes for sale at the school. A recent tourist venture, undertaken by the Women's Committee, promoted the beaches. Small *fales* have been erected and an ablution block was nearing completion (December 1986). Pre-booked tour buses stopped at the village and the Women's Committee prepared a Samoan lunch and entertained the tourists. There was evidence in almost 59 per cent of the homes visited, that a household member had been overseas or that goods had been sent from abroad. Despite the fact that there was no electricity, one house had a complete video set 'waiting'. Yet, many of the other village houses had very few material possessions. Apart from these few local businesses and evidence of remittance income, agriculture was the major source of revenue.

In early discussions, it was found that the women did have a 'set' of responses concerning the agricultural roles of women: 'agriculture is men's work, we hardly do that'. The common responses regarding women's jobs were:

- 6–10 year olds collect nuts;
- 11–15 year olds collect taro and nuts;
- 16–20 year olds weed, weave baskets, and prepare the foods for market, and do the copra (collect nuts, split nuts, scoop out meat, dry meat);
- 20 years and older are responsible for vegetables, weeding the home garden, however, older women do the household tasks and do not go into the garden.

As interviews and observations progressed, it became obvious, however, that women were playing a substantial role in agriculture and contrary to public assumptions, women were going to the plantations.

Village land

As in other Samoan villages, there were three discernible land-use zones in the village. Around the village *fales* (houses) food crops were grown

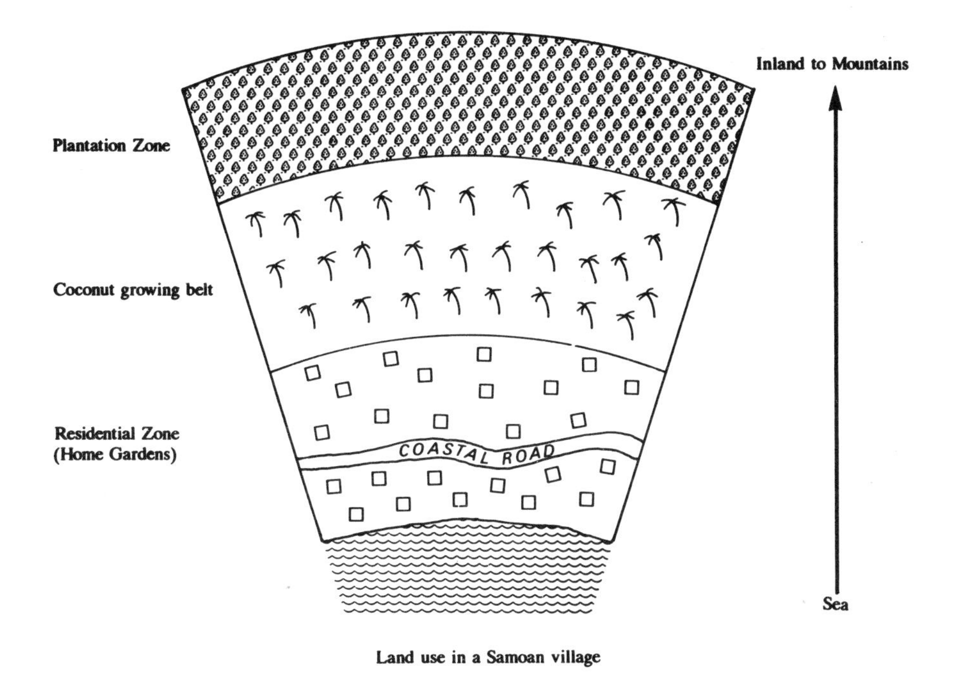

Figure 16.1 Land use in a Samoan village

mainly for household use, and there was some *laufala* (weaving material). Beyond the residential zone, was the coconut growing belt. A German ordinance of the early 1900s decreed that every Samoan landholder must plant 50 coconut plants in this belt annually, thereby pushing back the main crop-growing zone into the hills (Figure 16.1).

The pressure of land had led to the opening up of the hinterlands, with the result that many newer plantation blocks were a great distance from the village homesteads. No accurate estimation of the distances between home and plantation was made, however, the following comments give some indication of distance: 'plantations are 7 km away', '2 hours by horse', 'we have to catch a bus to the top, and walk back in'.

Every village family has access to land in each of the three zones. Plantation blocks varied in size and numbers, but were usually small blocks, and seldom contiguous.

Findings

(1) The main aim of the household was to provide sufficient food for family needs. Of secondary importance was the production of surplus crops for sale to enable the families to purchase items for household use

(such as blinds, mats, brooms, and oils) and other goods such as *ietoga* (fine mats) which are necessary for ceremonies and cultural exchanges.

Although the respondents said that food was of prime importance, the impression gained was that the need for cash was so strong, that households would probably sell foods they might otherwise have eaten. Cash was needed to buy oil for lamps, insecticides, fertilisers, foodstuffs such as tinned fish, sugar, tea, salt and rice, and for services, such as school fees, doctors and medicines, and church donations. During the week of the study, no vegetables were served at meals, and there appeared to be malnutrition among the children.

(2) Agricultural production was a daily, highly organised activity in all households. Women played a major role in the family work force in both the growing and preparation of goods for home use and market, and the selling of these. The amount and type of participation was partly determined by the composition of the household (sex, age) and the rank of the family, but the bottom line was that 'all labour is needed'.

Household gardens and animal care

Forty-three of the 54 households grew food crops around the homestead. It was not determined why the other 11 households did not have gardens, but it was noted that these households were all in one section of the village.

All houses had *laufala*, (pandanus weaving material) patches which were carefully tended. The women and girls cut, cleaned, dried and prepared large strips of *laufala*, winding these into rolls, most of which would be used by the household, but some would be sold at the market.

As expected, women were mainly responsible for the care and upkeep of the household gardens and of the free-ranging chickens and pigs. Sometimes youths helped with these tasks, particularly with the preparation of the gardens for planting. All land areas fronting the road were exceptionally well kept and the trimming of grass and sweeping of leaves was a daily duty for the women and children.

Coconut areas

Coconut-collection areas were generally between 5 and 15 minutes walk from the residential areas. Collecting nuts for household use and for copra making was the work of the women and children, to be done 'at any time', however, the majority of copra production was done by the young women. Copra making was the task that 'is always there; if we need money we can do'. Therefore, although copra prices were low at the time of the study, women still engaged in this activity.

No maintenance work was carried out in the coconut belt during the study and it appeared that little maintenance was necessary at any time.

Plantations

The most significant finding of this study was the extent to which women reported they were going to the plantations. In three families, 'mainly men' went to the plantation, and in three families, 'mainly women'. In the remainder of the families, plantation work was shared between the men and women although they might do different tasks. Actual time spent on the plantation was not measured, but one woman's comment summed up those of many other respondents: 'Most of us have to go . . . it is necessary . . . our only money is from the plantation.'

Plantations were long distances from the home site. This not only brought added problems of transportation, but increased the need to visit the plantations regularly to guard against theft.

Fishing

Women did the reef and shallow-water fishing. At least one woman from each household went fishing 'when the tides were right'. Catches were not as great as in the past. However, fish was a valuable addition to the diet.

Marketing

Eighty-three per cent of the families sold goods regularly at the Apia markets. Women helped prepare the crops for market and they plaited the carrying baskets. Women also did the majority of the market selling. Women in the 30 to 40 age range were said to be 'the best sellers', because they 'know how to handle money'.

(3) Traditional crops (taro and other root crops, banana) were the major produce grown. The people knew how to grow these and cash returns were reliable. The women expressed a strong interest in vegetable growing, but economic and information constraints affected this (Table 16.1).

It was found that vegetables were being grown on the plantation which could be one reason why women were going more frequently to the plantation areas.

(4) The main constraints to agricultural production were: not enough agricultural information and support; distance of plantation areas from village and lack of transport, both to the plantations and to the market.

Women do not feature in the formulation of the Rural Development project programme. The committee is chaired by the Prime Minister

Table 16.1 Reasons vegetable gardens were unsuccessful

	Respondents
Pigs, hen damage	14
Insects (lack of pesticides)	13
Soil no good	3
Seeds (hard to get, expensive)	3
Not enough money for labour	3
Climate too wet	2

Source: Author's fieldwork

and is comprised of the Ministers and Directors of Agricultural and Economic Affairs, the Financial Secretary, the Managers of the Development Bank and Agricultural Store, the Dean of the University of the South Pacific School of Agriculture, the Secretary of the Pulenuu Committee, the Secretary of Youth, Sports and Cultural Affairs and the Senior Rural Development Officer. Nor, as seen in Figure 16.2, do women feature significantly in the official line of communication in the Rural Development unit.

As stated, it was government policy that all village information be directed through the Pulenuu network of the Rural Development Department. The Pulenuu then passed on information to the Meeting of Chiefs (*fono o matai*) and finally the individual *matai* then passed information to his family.

Thomas (1981) suggests there are obvious shortcomings in this chain of communication, relating firstly to the distance between policy formulation, information and user, and secondly to the reliance on the ability and interest of the *pulenuu* to understand and pass on relevant and accurate information. Thomas cites an extension officer's comment:

> We don't use the *pulenuu* much . . . we can't explain technical terms to him and expect him to remember and pass information on correctly. Too much depends on the quality of the *pulenuu*. Many have no authority.
>
> (1981: 39).

Not only were official government communications made through the village *fono* network, it was also the cultural expectation that any approaches into the village must be made through the village council of *matai*. Therefore, women's access to informal sources of information was second hand.

Although there are no official policies indicating that agricultural information should be directed to men rather than women, established communication habits mean that the extension officers (all male) tended to talk to men. In this study, women tended to 'find information where we can' (Table 16.2).

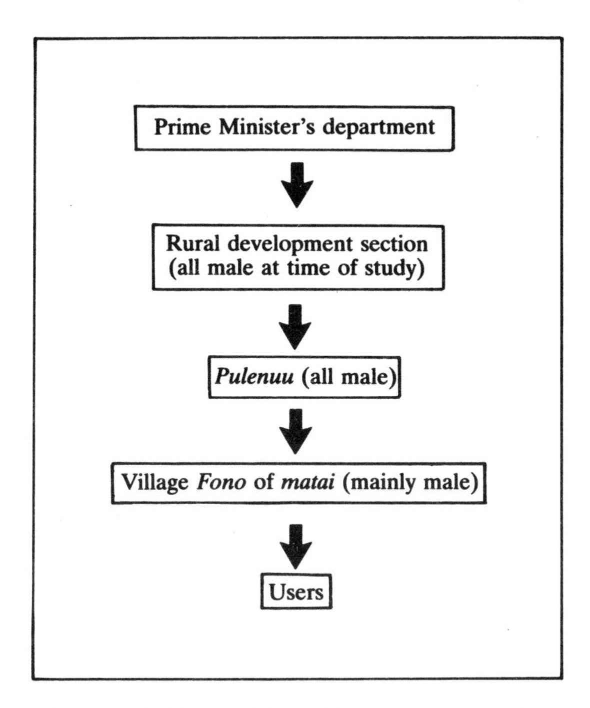

Figure 16.2 Hierarchical village communication

Table 16.2 Agricultural information sources (women's perception: n = 48)

Traditional, always done it like that	17
Personal experience, watching others	8
Agricultural department	8
Women's committee	6
School teachers	4
Talking with other farmers	3
Extension officer	2

Source: Author's fieldwork

Seven respondents stated that they wished the extension officer would come more regularly to the village. It is interesting to note that women preferred to go straight to the Agricultural Department in Apia than to wait in the village for an opportunity to talk to an extension officer. Although the church was not mentioned as a source of agricultural information, it was observed that the Catholic and Mormon churches both had 'model' back-yard gardens, which were fenced and laid out to grow both old and new crops. The 'neighbour' effect was also evident in the village; if one house had a certain type of vegetable in the garden, it was likely that the neighbouring house would also.

The distance of plantations from homesteads has already been noted. Much time and energy was expended walking to and from the plantation areas and carrying produce back to the house. The transport of goods to the market raised problems of cost and availability of transport. Goods were taken by the early morning bus to Apia, or by hired car; both costly. As there are no storage facilities in the village, bags of crops were often 'squeezed' on to the bus, became bruised and thereby devalued.

(5) The market potential for handicrafts has been realised and this affected production behaviour. In most homes the elderly women weaved for most of the day, middle-aged women for parts of the day and the other women weaved in their spare time. The women made essential household items, items for ceremonial use, and a surplus of these for sale. Items for the pastor's house and church were made at the weekly group weaving days (*falelalaga*) held in the pastor's house.

It is only very recently that *ie toga* (fine ceremonial mats) have become a marketable commodity; bought by an urban population who do not have the time to weave their own, or exported to Samoans living abroad and maintaining traditional ceremonies. Mat making has always been regarded as an art, with the making of the fine mat the most prized artistic piece. Each fine lauie thread was painstakingly prepared and placed; the finished product resembling a delicate linen which 'ruffled in the wind'. Each completed mat was a reflection of its weaver, and ownership was acknowledged in reciprocal exchanges. The new *ie toga* are made from *laufala*, a thicker, coarser fibre which is quicker and easier to weave. The weft is coarser; the care is less; the pride is minimal; for if a product is for sale, who is to know of its origin? This new attitude to weaving was apparent in this village. Although some said that they never sold their mats, particularly the *ie toga*, others said they did when they needed the money. Mat making was seen in every house, and differences in quality of product were noted. In some houses, two women sat weaving the same mat. One woman said that she could finish one *ie toga* in one day. This was in marked contrast to her elderly aunt who sat carefully scanning each centimetre of mat for irregularities, undoing rows which did not meet her standard.

(6) Women are looking for alternative ways of earning revenue. They are, therefore, potential adopters of new cash-generating technologies.

CONCLUSION

Women in this village were making a significant contribution to the household economy, producing food crops and goods necessary for home use, ceremonial and sale. The need for revenue had affected the production behaviour in each of the households. Not only were women trying to produce more handicrafts they were also 'doing more agriculture'.

Agricultural production is becoming a major way in which cash incomes can be gained today. Thomas and Hill, for example, state that, given development priorities and the current economic situation, the most accessible form of income generation for most Pacific Island women is likely to be in informal marketing or agriculture (1986: 119). The increasing complexities associated with growing crops for local and overseas markets (quality control, timing of production, packaging, transport) makes access to technical knowledge and assistance vital.

Women's increased contribution to smallholder production has not been acknowledged, therefore this contribution is not considered in agricultural policy, nor are women given automatic access to information-support systems which will enable them to increase their efficiency.

Agricultural planning should be based on women's actual contribution to agricultural production, rather than public perceptions of this role. Firstly, more accurate statistics are necessary. Secondly, more effective linkages must be established between rural women and useful planning agencies. This should lead to a utilisation of the potential contribution of women to developing food sufficiency, better-quality exports and an overall improvement in quality of life.

Part V

LATIN AMERICA

INTRODUCTION

Gender and geography in Latin America

A historical perspective

Janet Momsen

Despite geography existing as a discipline in most Latin American universities and attracting a higher proportion of women faculty than in other parts of the world, geographical studies of gender have been slow to develop. This apparent paradox may be related to geography's strategic link with the military in many Latin American states and the political barriers to the introduction of radical ideas in universities until recently. In addition, although Latin America has a thriving broad-based feminist movement holding regional conferences since 1981, it is a very diverse movement.

The ideological and strategic discourses of Latin American feminism have focused on two relationships: that between feminism and the revolutionary struggle for justice and that between what was a predominantly middle-class feminist movement and the growing popular-based women's movements. (Sternbach *et al.* 1992: 432). Brazil has not only the largest, most radical and most politically influential of all Latin American feminist movements but also has the most geographers. Yet when I organised a meeting on women and agriculture in north-east Brazil in 1982, the only Latin American speakers were Brazilian sociologists and agricultural economists. Geographical research with a gender focus in the region has continued to be carried out predominantly by European and American geographers as the chapters in this volume illustrate.

When I spoke on gender and geography at an Anglo-Mexican conference in 1985 (Momsen 1986), the most positive responses came from young graduate students. This cohort now dominates the work in this field by Latin American and Caribbean geographers (Calio 1991; Franco 1991; Tadeo 1991; and Harry 1992). In 1991, when Rosa Rossini, who wrote the first doctoral dissertation on gender and geography in Brazil (Rossini 1989), organised a conference in São Paulo on behalf of the International Geographical Union Gender Study Group, eight papers were presented by Argentine, Brazilian and Mexican geographers. Many of the authors of these papers are activists as well as academics and they chose to report predominantly on urban and rural social movements. They thus are providing a new and radical voice in Latin American geography.

Women and geography in Brazil

Sonia Alves Calio

The dictionary of feminist ideology by Victoria Sau very clearly defines what can be called the androcentrism of geography: 'Man as the measure of all things. Studies and research done uniquely from the masculine point of view and the subsequent application of the results to the ensemble of individual men and women.'

Geographical research in Brazil has not avoided this androcentrism: women are excluded as social subjects whether in studies of broad social movements or in analyses linked to the capitalist organisation of Brazilian space. One has been able to see this situation for oneself by surveying the various reports on the Congresses and Meetings of Brazilian geographers which have taken place these last five years. At these meetings absolutely no papers have been presented on gender. [When the IGU met in Brazil in 1982, the Rural Commission held a session on women and agriculture organised by Janet Momsen. The speakers at this meeting included Anglo-American geographers and Brazilian sociologists but no Brazilian geographers, although Brazilian geographers were present. (Ed.)].

Unfortunately, geography in Brazil has not followed the academic debate which has recently been established in the nation's sociology, concerning the articulation of social class and gender. This debate is the fruit of the active existence of an important women's movement in Brazil since 1975. Groups of women, feminists or not; publications; student groups; research commissions; all of these have appeared and continued to appear and participate in this debate. Nonetheless, this is not the case for the academic life of Brazilian geography: there are no courses at university, no scientific commissions, no interdisciplinary research. Even today, nothing is proposed.

However, two isolated pioneer examples must be emphasised. First, the Department of Geography of the University of São Paulo has, since 1989, offered, within the official syllabus of the programme Third Cycle, a course entitled 'The woman as labour force in agriculture', proposed by the geographer Rosa Ester Rossini, who has for some time been carrying out research on women and agriculture. In 1989, Dr

Rossini wrote the first geography doctoral thesis in Brazil on a topic concerning gender. Secondly, I myself have just completed with the same Department of Geography, a doctoral thesis on the topic 'Urban structure and the spatial segregation of women'. Supported by my feminism and my activity in the feminist movement, I have studied questions concerning the relationships between patriarchy and capitalism in the urban system, the gender division of labour and the gender relations of men and women.

These achievements mark the first steps in feminist geography in Brazil. They are timid, solitary steps, still bereft of theoretical unity. However, the fact that we have done so little in travelling this road must not obscure the importance of our role in rethinking the question of the androcentric geography of Brazil.

It has led us to take an activist and policy-making direction. For example, our work has begun to bear fruit not directly at the level of the university but at the level of one of the nation's municipalities which is in the process of preparing its master plan. The Prefecture of Santo Andre, an industrial municipality of the State of São Paulo, administered by a left-wing party (the Workers Party), has given us a research contract which involves considering the city and urban reform from the point of view of women and helping women to participate in the process of preparing the municipal master plan. Through public meetings, women have been brought together, have been listened to, and have made joint trips in the city with the common aim of 'thinking about the city in the feminine'. The results are surprising: the women have realised that the city does not belong to them, and that if they wish to be full citizens they must appropriate it.

According to the national constitution, the master plan is compulsory for all municipalities with more than 20,000 inhabitants and had to be approved by the end of March 1991 by the electorate and by the local councillors. With the assistance of urban geographers, the master plan of Santo Andre will be the first in the country to have the participation of women. This is a unique experience in a country that has always been seen as masculine.

The city of São Paulo itself now has a woman mayor, Luiza Erundina, who is also a member of the Workers Party. She has set up a Woman's Secretariat and is very sensitive to the question of women in the city. This Secretariat has learned of our work in Santo Andre and we have been invited to discussions in the Department of Urban Planning. Thus, we are becoming both academic and applied feminist geographers.

17

THE ROLE OF WOMEN'S ORGANISATIONS AND GROUPS IN COMMUNITY DEVELOPMENT

A case study of Bolivia

Jane Benton

INTRODUCTION

This paper draws on earlier investigations of the changing lives of marketing women from the Lake Titicaca region and is based on a 1987 research project aimed at evaluating community development programmes run by Bolivian non-governmental organisations (NGOs) with overseas financial and/or technical aid[1] (Figure 17.1). Although it was not my original intention to focus on gender issues, it became apparent at the outset that a growing number of Bolivian NGOs are being established and staffed entirely by women to work exclusively with women's groups in *barrios* (low-income urban areas). Additionally, a number of male-controlled NGOs have decided over the years to concentrate their efforts on women's groups in rural and urban areas; others operate separate community development programmes specifically for women. Simultaneously, some women's grassroots groups are taking matters into their own hands and are currently engaged in activities calculated to bring about beneficial changes in the lives of their families and communities.

In *Community Participation, Social Development and the State* (1986) Midgley *et al.* discuss the respective contributions made by state governments and non-governmental organisations to community development in rural areas of the third world. They claim that most of the successful programmes using community participation as a principal method of operation have been run by non-governmental organisations. Several reasons are advanced: NGOs are unencumbered by bureaucratic machinery, their projects are usually on a smaller scale, making community participation easier to achieve; because of their charitable nature they are more likely to employ highly motivated people unconcerned

Figure 17.1 Bolivia: Location of non-government organisations visited in this study

about career prospects; and they find it easier to be innovative and initiate experimental projects that demonstrate a potential usually ignored by bureaucrats.

Sadly, any such theorising about the relative merits of state or non-state intervention at community level is almost irrelevant in connection with present-day Bolivia. Due to the country's grave economic situation, NGOs are obliged to operate in a vacuum, attempting to satisfy the often-desperate needs of people suffering as a result of circumstances – such as the economic recession and the collapse of the tin market – totally beyond their control and comprehension. While the central government presents no challenge and places few impediments in the way of NGOs, state development corporations, severely handicapped by lack of funds and faced by formidable gaps and deficiencies in the provision of housing, health facilities, sanitation, education etc., openly welcome opportunities to collaborate with national NGOs and international organisations such as UNICEF and CARE.

Over the last two decades and particularly since the 1982–3 drought, which affected extensive areas of Bolivia, countless NGOs have emerged; as elsewhere in Latin America, some have been short lived but others have taken root and achieved varying degrees of success. Despite the praiseworthy efforts of bodies such as the Centre of Documentation and Information (CEDOIN), who collate data about national NGOs, there is no agreement on the number of organisations currently working in the country, though it must exceed 500. Many of the NGOs are Roman Catholic institutions; others could be categorised as main-stream Protestant (the Methodist church has been involved in education, health and agricultural programmes since 1906) while hundreds of organisations have no religious affiliations. According to Sandoval (1987), Bolivian NGOs have been created by 'heterogeneous sectors – religious, political, humanist, academic, union, professional and co-operative.'

In 1986 the Centre of Information and Women's Development (CIDEM), a feminist NGO, published a directory of institutions and organisations that 'work on behalf of the women in the cities of La Paz, Sucre, Cochabamba, Santa Cruz, Potosi and Oruro'.[2] The book contains 95 entries, although several organisations such as CARITAS are mentioned more than once, and five references are made to radio stations broadcasting programmes for women on health, hygiene, nutrition, legal rights, adult literacy and 'popular education'.[3] Whatever the exact number of such organisations, it is substantial and growing. Simultaneously, the number of women's grassroots groups has increased significantly since the early 1980s. The varied nature of such groups is best illustrated by referring to the work of the Centre for Self-Development (CDA), an organisation run by a couple from Argentina. Adult literacy is the prime concern of CDA. In 1987, the organisation was working with 66 women's groups (ranging in size from 50–300), falling into five categories – housewives' committees/mothers' clubs, groups of factory workers, market syndicates, *campesina* (peasant) groups and the wives of relocated miners.

ECONOMIC BACKGROUND

The present government claims to be deeply concerned about living conditions in rural areas and the ever-expanding *barrios* encircling most Bolivian cities. The government attributes its present lack of activity to the economic crisis, that is, scarcity of cash resources for investment in small-scale development planning. Whilst this cannot be refuted, there is little evidence of such a commitment on the part of central government over the last twenty years. Since the introduction of the New Economic Policy (Decreto 21060 was presented to the nation on 29 August 1985) the lot of the *campesino* (peasant) and rural–urban migrant has deteriorated considerably. In an effort to reduce hyper-inflation (then at a world record of 24,000 per cent),[4] to cut public spending, increase exports and reduce imports by devaluation of the currency and reduction in domestic demand, wages in the public sector were frozen, subsidies on bread, electricity and transport were removed and VAT was put on essential medicines and school supplies. In August 1988, Vacaflor observed:

> Since the collapse of the tin market in late 1985 the payroll of Comibol [the state mining corporation] has shrunk from 31,000 to 6,000 . . . unemployment, further aggravated by public spending cuts, has reached 25 per cent, according to unofficial figures.

The rural sector has fared equally badly. Over the last few years crop diseases (especially in potatoes), prolonged drought and the ready availability of US Aid in the form of food donations have seriously demoralised the farming population: by the mid 1980s Bolivia had ceased to be self-sufficient in commodities such as wheat. In a series of articles in *Presencia* in August 1987, Luis Antezana Ergueta, Director of the National Council for Agrarian Reform, expressed deep disappointment and concern that, while the area under cultivation in the departments of Santa Cruz and Tarija had increased by 150,000 hectares since the introduction of agrarian reform in 1953, rural–urban migration on the *altiplano* had led to the abandonment of 200,000 hectares, a 30 per cent decrease. Over the country as a whole there had been no increase in the area devoted to crops; in fact, in many parts output had decreased substantially and in some regions soil erosion had reached alarming proportions. Meantime, rural–urban migration, draining the countryside of young, energetic people with progressive ideas, had accelerated to such an extent that by the mid 1980s El Alto, the area of *barrios* overlooking La Paz, had grown to half the size of the city proper, and was petitioning for separate city status.

MACHISMO AND WOMEN'S TRADITIONAL ROLES IN BOLIVIAN SOCIETY

The strength of the women's movement in Bolivia may appear surprising; the very emergence of organised women's groups in a country where *machismo* continues to flourish within the patriarchal structure of the extended family is no mean feat. As Harris (1983) reminds us: 'The proper characteristics of a [Latin American] woman are . . . submissiveness, dependence, exclusive devotion to family and home'. Figueroa and Anderson (1981) state that:

> women are socialised to respond to the wishes of others, not to define clearly their own desires. They are socialised to think of themselves as less worthy of men. Even in rural areas . . . a midwife is paid more if the baby she delivers is a boy.

The Bolivian *campesina* has traditionally endured extremely arduous toil in the fields (often with a baby on her back), combined with abysmal poverty and physical violence in the home. With few exceptions those *campesinas* interviewed in 1971 were withdrawn and submissive: frequent rejoinders were 'only my husband can answer for me' and 'my husband will beat me if he finds out that I've been talking to you'. Even though generations of women have made a significant contribution to agricultural production, rarely, except in the case of some widows, have they been recognised as members of agrarian co-operatives or allowed to participate in community decision making. Frequently the only alternative to a life of constant struggle in the countryside has been domestic service in the city, with its concomitant economic and sexual exploitation.

As rural–urban migrants, Quechua and Aymara women face other difficulties: 'We peasant women, we are discriminated against from the moment we leave home when we go to the city or set foot in the village. They call us Indians, peasants, filth, and all that. These are problems that we have to face' (Margarita Laime)[5]. Many such women work either in the 'informal' sector, in which case their work is seen by men as an extension of household duties and viewed as non-productive, or in the least skilled jobs (often in appalling conditions, for very low wages and no job security), which they are obliged to take in order to combine waged labour with household responsibilities. According to a recent study by the UN Commission on Latin America and the Caribbean such exploitation is now widespread in Latin America as a result of 'the region's economic crisis . . . the drop in family income has driven young women into the workforce, but no similar apparatus is working to ensure their effective absorption into the economy . . . In 1985 women earned only 54–84 per cent of what their male counterparts earned.'[6]

As elsewhere in Latin America, the feminist movement in Bolivia has gathered strength in the last decade as a direct response to the abuses and perpetuation of *machismo*. Examples of women bringing divorce cases (often with the support of female legal advisers) against unfaithful and physically abusive men are no longer rare. While abortion remains illegal and the government, with its pro-natalist policy, provides no support for family planning, middle-class women are able to obtain various forms of contraception and thus have greater control over their reproductive systems.[7] An ever-increasing number of women are gaining access to higher education and entering the professions. For example, half of Bolivia's teachers are women; they earn the same salaries as male teachers and 'have repeatedly demonstrated their combativity in wage demands' (Latin America and Caribbean Women's Collective 1980).

Gott (1987) refers disparagingly to a 'regular feminist programme' on Bolivian television:

> It shows well-dressed, well-educated, attractive but slightly stupid young women from the upper classes going out to interview women who are older, uglier, poorer and more intelligent than they are about the problems women face in contemporary Bolivian society.

Such 'slightly stupid' upper-class women are very far removed from the intelligent, well-educated, middle-class professional women working in the NGOs visited in 1987. The latter are the first to appreciate the problems created by the enormous cultural gap between themselves and the women with whom they want to work. A CIDEM handbook (distributed in 1987) highlights the project workers' dilemma about how to establish contact with *barrio* women:

> What ought to be the form of communication with these groups? What language should we use? All of us were women, but sadly there were great differences . . . We were women who had access to study and professional training. The groups with which we were proposing to work were for the most part composed of women with an experience of life far fuller than our own . . . all of it a road passing through the streets of misery. They had a different view of the world. They were groups that had a totally different cultural universe.
>
> (Alzerreca and Ruiz 1987: 7)

CHANGING ROLES OF PEASANT WOMEN AND RURAL–URBAN MIGRANTS

While the majority of *campesinas* living in isolated communities continue to endure untold hardship, often burdened with heavy workloads that threaten their life and well-being, the lifestyles of some more fortunate Aymara and Quechua women have changed dramatically over recent years, a fact that helps to account for the proliferation of women's groups. Access to secondary education has for some resulted in an increase in social and political awareness and heightened aspirations, self-assurance and assertiveness. Even by 1981 a number of *campesinas* interviewed in the Lake Titicaca region had become highly critical of the freedom enjoyed by men without domestic responsibilities; they had for some years been demanding the right to attend community meetings and participate in decision making regarding matters likely to have a direct effect on their lives.

As the pace of rural–urban migration accelerates, more and more campesinas are moving to urban environments, either with their families or in some cases for educational and professional training purposes. For example, a significant number of women born in the lakeside region have now entered the teaching, nursing and police professions. Others have improved their entrepreneurial potential by becoming fluent in Spanish and learning simple accountancy; some of the most successful marketing women travel widely and freely in the course of their work and may wield considerable power and influence as leaders of market syndicates.

Additionally, an ever-increasing number of women are obliged to make family decisions and assume financial responsibility for their households because of illness or the death of their spouse, desertion or the temporary migration of family members to seek work elsewhere. These and other factors have led women to examine their traditional roles and consider new, often alien, survival strategies in terms of income generation. To such women, membership in a mothers' club or housewives' committee bestows a feeling of belonging and sense of security; such groups promote unity and strength of purpose to achieve set goals, such as the installation of a drinking-water supply or the procurement of a teacher to initiate adult literacy classes.

Women's unions

Undoubtedly, the militancy of the tin miner's wives has made a deep impression on the women's movement in general. The Housewives' Committee of the Siglo XX mine was established as early as 1961 under the formidable leadership of Domitila Barrios de Chungara, later to become a spokeswoman at international conferences on behalf of all marginalised Latin American women. In its campaign for miners' rights, the committee has frequently challenged authority; in 1978, the committee began a hunger strike which eventually led to the downfall of the national government. In 1985, as part of a campaign against the government's newly imposed austerity measures, the miners' wives initiated an equally memorable hunger strike, eventually involving themselves and 10,000 miner husbands. In the following August, these same women participated in the miners' ill-fated 'March for Life and Peace'.

In the words of Domitila:

> We women were raised from the cradle with the idea that women were made only to cook and take care of the kids, that we are incapable of assuming important tasks and that we shouldn't be allowed to get involved in politics. But necessity makes us change our lives . . . Since [1961] this committee has always been in step with the unions and other working-class organisations, struggling for the same causes.'
>
> (Barrios de Chungara and Viezzer 1978)

Bronstein (1982) records the comments of Rosa Dominga, one of the *campesina* organisers of the First National Congress of Bolivian Peasant Women held in La Paz in 1980.

> Women are beginning to discuss, to ask questions, to talk about their needs. If we work together maybe we can even make the government listen to us. But if we stay in our houses we have nothing. No one is going to take any notice of us. If we are organised, if we can come up with resolutions, we can publish them in newspapers, on the radio, so everyone will know what we are thinking. Everyone will begin to think about the *campesina* woman. I have seen how the factory workers and the miners organise, and we are equal. We are producers and workers too. Why shouldn't we have our own organization to fight for our own needs?[8]

Despite the fact that few *campesinas* had previously been allowed to attend their own community meetings, let alone make suggestions and vote, over 1,000 women travelled to the congress to discuss a variety of issues including health, education and marketing facilities. Delegates issued a 64-point manifesto advocating improvements in their political, social, economic, educational, and cultural situation; as a result of their deliberations the National Federation of Bolivian Peasant Women (FNMCB) came into being, affiliated to the national Trades Union Congress (COB); a remarkable achievement.

It is apparent from these examples that marginalised women, when united in strength and purpose, can be a powerful and effective force for change. Some Quechua women attribute their belligerency and tenacity to their Inca ancestors. Aymara women say that their people have engaged in struggle after struggle since their subjugation by Inca overlords. As a result they have never had anything to lose and have acquired a spirit of perseverance and determination. Bolivia has a long and impressive history of courageous Indian women. Such heroines continue to inspire women; one *barrio* women's group working with the Bolivian Centre for Educational Investigation and Action decided to name their women's centre after Bartolina Sisa, who, in the struggle for the independence of Upper Peru, organised the siege of La Paz in 1781.

Women's participation in grassroots organisations

A number of Bolivian NGOs work directly with housewives' committees and mothers' clubs in shanty towns and mining areas. The former have a union-like structure and, in the case of the Siglo XX committee tend to attract energetic and militant members who are committed to change. They are not feminist organisations as Domitila makes clear: 'Our position is not like the feminists' position. For us, the most important thing is the participation of the *companero* and the *companera* together. Only that way will we be able to see better days, become better people and see more happiness for everyone.' Since the collapse of the tin mining industry at the end of 1985, housewives' committees in the mining areas have collaborated with national and international aid agencies in operating emergency food programmes – both distributing donated food and cooking meals in communal kitchens.

Many mothers' clubs have been formed at the instigation of local Catholic parish churches. While they are formed initially to draw together young mothers and provide support and instruction on child care, health and hygiene, they take on a variety of roles, such as raising money to sponsor fiestas, to purchase school books and equipment, cooking breakfast for school children and collecting subscriptions from families in the community for use in case of emergency.

Other grassroots groups are less formally structured and normally emerge once women decide to tackle particular issues, such as health issues or water problems. Women may decide to group together to put pressure on local authorities or in order to establish contact with NGOs. I came across a number of examples of women grouping together because they had seen the advantages to be gained from group formation in neighbouring communities. Thus, the *carpas solares* (solar greenhouses) constructed by women's groups in El Alto with the collaboration of the Centre for Educational Investigation and Action have become the envy of adjoining communities. Once the greenhouses produced their first crop of tomatoes, radishes and lettuce several women's groups formed in order to make direct approaches to CEBIAE for assistance.

'Education can challenge the subservient role of women, giving them wage-earning potential, confidence, status, independence and new possibilities' (World University Service undated). The vast majority of marginalised women in Bolivia, with the highest rate of female illiteracy in South America,[9] appreciate the extreme importance of education and are prepared to make great sacrifice in order to attend adult literacy classes. Domitila's words admirably reflect their feelings: 'the more ignorant we women are, the longer it will be before we liberate ourselves'. Reasons for wanting to be able to read and write are numerous. They include a desire to be able to read official documents, to read letters from and write letters to children living away from home, to understand prices and shop signs in the markets, and to read bus directions and simple informative booklets on issues such as ante-natal care and legal rights. According to CDA, a number of women's groups have been formed specifically for educational reasons. (One CDA adult literacy class observed in a *barrio* above La Paz was split into two groups; while the women outside were sitting in a circle learning how to write their names, the more advanced group was inside the local school, debating how to cajole their men into moving rubbish from the streets and waste areas in order to combat the spread of infection and disease).

The provision of portable water is a basic need uniting women. Van Wijk-Sijbesma (1986) claims that 'women's programmes and organisations have great potential to mobilise women to improve their own water supply and sanitation if they are not served by other programmes'. By 1981 *campesinas* in several communities in the Lake Titicaca region had succeeded in pressurising the men into petitioning the state development corporation to install a reliable water supply; at the time of my visit in 1987, CARE was installing simple sanitation systems in a number of communities. Such programmes now incorporate instruction on nutrition and hygiene; CARE requires each community to select a woman for training as a health promoter.

In situations where women are obliged to assume total financial responsibility for their families economic problems are at their most acute, especially where there are young children to nurture or women have no experience of entrepreneurial activities. In view of the recent economic austerity measures, it is not surprising that infant malnutrition has grown at alarming rates. According to the National Nutrition Institute, one half of Bolivian children suffer from some form of malnutrition (Foronda 1989). Women, desperate for cash to buy food and medicines, are often forced into piece work at home, without any means of ensuring a realistic wage, such as a union or co-operative might provide. Under such circumstances, groups of women have eagerly responded to any offers made by NGOs to provide alternative forms of income generation. The Centre for Promotion of the Woman, 'Gregoria Apaza', a feminist organisation supported by OXFAM and CAFOD, works with women's groups in the *barrios* of La Paz and has been particularly concerned with finding practical ways of supplementing family incomes. Its present imaginative projects include enterprise devoted to the manufacture of *mantels* (the coarse aprons worn by Indian women) for sale in local markets, a small jam factory employing women on a shift system and a popular laundry using appropriate technology. In one of the communities, 'Gregoria Apaza' has opened a school for women which offers courses on nutrition, practical subjects and nursery nursing. The organisation's nurseries run for children (8 months to 6 years) of factory women workers, are highly valued by women living away from their extended families, which would normally be expected to provide help with child care.

Women's projects run by women's organisations

Some NGO's established and staffed entirely by women openly identify themselves as feminist and, as such, are striving towards the creation of equality and the elimination of all forms of discrimination against women in what remains a male-dominated society. The staff of 'Gregora Apaza' describe themselves as a group of women who want to see equality between men and women and to end class exploitation. Women working with such organisations view their role as that of an intermediary, making *barrio* women aware of the unnecessary oppression they suffer, of the factors that obstruct their escape from poverty and of their political and legal rights; they aim to 'help women to help themselves' by communicating basic organisational and leadership skills and by mobilising women's groups to instigate projects likely to improve their socioeconomic conditions.

While CIDEM, unlike 'Gregoria Apaza', has not ventured into income-generating enterprises, its activities range widely. Staff members run weekly courses for housewives' committees and mothers' clubs on subjects such as 'women and the law' and 'women and health'. Advanced courses for group leaders address issues surrounding methods of working with women and the use of posters, popular theatre and radio in their campaign to create social and political awareness among non-affiliated groups.

The Centre for the Promotion and Training of Women (CEPROMU), does not claim to be a feminist organisation, but rather 'an organisation which has brought together women professionals, university teachers and students with the common aim of assisting Bolivian women from the most underprivileged areas. CEPOMU views its role as a catalyst 'enabling' women to identify their own needs and seek their own solutions. It originated in 1981 after nine Cochabama University lecturers had spent a year discussing the problems of urban women and decided to make a commitment to more practical involvement.

Significantly, the work of CIDEM, 'Gregoria Apaza' and CEPROMU is confined to women's groups in *barrios*. A number of factors help to explain their reluctance to become involved with *campesinas*: the cultural gap between middle-class, professional, urban-dwelling women and *campesinas* is wider than that between the former and rural–urban migrants accustomed to city living and for the most part better educated than rural women; few women in such organisations have agricultural training or any experience of a rural way of life and the physical distance between NGO offices and rural communities pose insurmountable problems in organising courses and maintaining regular contact.

CONCLUSION

While 'strong and autonomous local initiatives can develop over time to meet more needs and make use of more potential than external NGOs can ever expect to do' (OECD 1988), any actual evaluation of the contribution made by women's organisations and groups to community development or of individual project sustainability is fraught with difficulties. It is impossible to avoid an element of subjectivity in any assessment and that 'pre-selected variables often [have] little to do with what people themselves, the intended beneficiaries, thought was most meaningful about the project' (Salmen 1988).

Temporarily successful projects may collapse when conflict develops within a community or between women's groups and NGOs. The withdrawal of financial support or interest on the part of the organisation may also cause projects to founder. A flourishing organisation such as Q'antati appears to offer favourable prospects for sustainable development but this depends on a number of factors over which the organisation may have little control: continued demand from overseas trading associations, ability to experiment with different products as markets for certain commodities become saturated or customer preferences change, and international exchange rates.

For many marginalised women, joining a women's group has made an unprecedented difference to their everyday lives of hardship and toil. Attending meetings provides a welcome diversion from a life of drudgery and, in many cases, of physical violence. They obtain a feeling of camaraderie from the group membership and have the opportunity to discuss problems and work together.

In the opinion of many of those interviewed, although adequate housing, water supplies and cash to buy food, clothing and medicines, are essential to life and any means of increasing household earnings are worthy of consideration, their greatest need is for education in the full sense of the term. Once literate, an individual cannot be deprived of the ability to read and write and countless opportunities are available. 'Popular education' programmes help women to recognise ways in which they are oppressed and to challenge the worst aspects of their exploitation; the self-confidence and management skills gained from group organisation and leadership have a life-long duration. As Swift (1988) maintains:

Education is an essential component of grassroots development. In Latin America it is called conscientization . . . whatever the label, it translates as empowerment – giving people the tools, not just to solve one immediate crisis but to analyze and respond to new and emerging problems.

18

DECONSTRUCTING THE HOUSEHOLD

Women's roles under commodity relations in highland Bolivia

Colin Sage

INTRODUCTION

Feminist analysis has long regarded the household as the site of domination and subordination and warned of the dangers of characterising households by reference to virtues of pooling, sharing and generosity (Harris 1981). More recently, other critical approaches have begun to recognise the importance of connecting the internal structure and social relations of the household to the development and intensification of commodity relations (Friedmann 1986). Yet, despite a greater sensitivity to changing gender and other divisions of labour based on age and marital status within the household, attention to individual activities and aspirations frequently remains subsumed to a preoccupation with the characteristics of the collective unit.

Writers on commoditisation have focused upon the process of deepening commodity relations within the reproductive cycle of the household. This process is illustrated by the household's increasing reliance upon transactions conducted through the market for the sale of produce and labour power, and for the purchase of necessary consumption goods and renewal of means of production (Bernstein 1979). Commoditisation leads to the 'individualisation' of the household as direct reciprocal ties between units are replaced by market relations and households enter into increasing competition in their efforts to increase control over land, labour and other means of production (Friedmann 1980). While the commoditisation approach is a valuable tool for analysing the ways in which households as units of production are incorporated into the market and subordinated to capital, it contains serious limitations in its ability to reveal the changing roles and circumstances within the domestic unit. In particular, the model fails to give sufficient attention to the persistence of non-commoditised (i.e.

reciprocal and co-operative) labour relations, the non-monetised exchange of goods and the continued importance of use-value production (Long 1986).

A view of the household that commences from a recognition of the unequal distribution of power and access to resources is better placed to examine the roles, relationships and activities of its individual members. This serves to 'deconstruct' the household, moving beyond the 'black box' conception of a unit collectively engaged in a single form of production (Whitehead 1981). Rather, the household serves as a locus for supporting simultaneous involvement in various complementary or non-articulated spheres of production. These are likely to include both agricultural and non-agricultural activities. However, the precise combination and weight of such activities will be influenced by the level of commodity relations and their impact upon the internal structure of the household, as well as its resource base of land, livestock, capital and labour. This will ultimately determine the nature of the household's insertion into the local economy and, consequently, its position with regard to processes of differentiation. For this reason, although it is necessary to deconstruct social relations *within* the household, it remains vital to establish the access of each household to the means of production.

The purpose of this chapter is to explore the role and importance of women's economic activities in relation to the organisational structure and resource base of the household. Such activities can only be fully understood against the background of women's reproductive responsibilities. Data relating to biological reproduction in Santa Rosa, Bolivia, are presented to underline the burden of child bearing and rearing for women's well-being. This is followed by an examination of the two principal economic activities in which women in Santa Rosa are involved: petty trading and brewing. The chapter concludes with two case studies which illustrate the importance of understanding the dynamics of individual lives and the range of women's responses in dealing with their personal situations.

AGRARIAN CHANGE IN SANTA ROSA

Santa Rosa lies approximately 100 kilometres north-west of the city of Cochabamba in Bolivia. The village sits roughly mid-way up a hillside that stretches from the semi-tropical valley floor below 2,000 metres to the temperate *puna* land at over 4,000 metres above sea level. Access into the area is difficult because of the extremely rugged terrain; the single road crosses a pass at over 4,200 metres but at lower altitudes is frequently blocked by landslides or washed away by rivers during the rainy season. Restoring the direct road link with Cochabamba on an annual basis is hampered by a lack of state investment, which limits infrastructural development in the region. The state plays a largely peripheral role, with a nominal presence that serves to legitimate networks of local power exercised by intermediaries in alliance with large local landowners and residents of the larger rural settlements. Transport into and within the area is provided by independent truck owner-operators who monopolise the movement of goods and people. Cultural distinctions reinforce the economic, social and political subordination of the largely monolingual *quechua* speakers of the mountains to the Spanish speaking rural elite (Dandler 1971).

It is under this historically oppressive, hierarchical social structure and situation of physical marginality that household production has become incorporated into the Cochabamba regional economy, a process invigorated by the implementation of agrarian reform in the early 1960s. In contrast to the heavily restricted economic opportunities under the semi-feudal *hacienda* regime, most households in Santa Rosa today are involved in the relatively specialised production of potatoes for the regional market.

While the involvement of women in agricultural production varies according to landholding strata, family lifecycle and household composition, it is clear that the contemporary sexual division of labour is more strictly defined than under the *hacienda* regime. Although women are allocated such tasks as sowing, with all its associations with fertility, they are rigidly excluded from ploughing. Thus, men are able to carry out all stages of the agricultural cycle, including the planting of seed when necessary, while women – even in their role as head of household – are sanctioned to rely upon men for particular tasks (Radcliffe 1986). This shift in the balance of gender relations has undoubtedly accompanied commoditisation (Bourque and Warren 1981).

Nevertheless, the limitations on capital accumulation from agricultural production, due to low producer prices, and the locally restricted opportunities for sufficiently remunerative waged employment, highlight the importance of non-agricultural, income-generating activities, many of which are controlled by women. These activities fulfil fundamentally different functions according to the characteristics and resource base of the household. In the case of those households headed by women who, without access to male labour and under dominant notions of gender roles, are prevented from fully participating in agricultural commodity production, such income-generating activities as petty trading and maize beer (*chicha*) production are central to household survival.

THE HOUSEHOLD: CONCEPTUAL CONSIDERATIONS

As the level of commodity relations deepens and horizontal ties between units are severed or transformed, social relations within the household may be despotically reinforced in order to control the labour power of family members (Smith 1986). Preventing the sub-division of land and maintaining control over the labour of adult sons and daughters is a source of considerable internal conflict with households, particularly among those with the greatest resources. Between parents and their children there can be substantial discord over the allocation of roles and distributional rewards, and this may be accompanied by struggles between siblings over proposed divisions of family property (Long 1977). The principal conflict in Santa Rosa, however, arises from the irreconcilable differences between the objectives of the senior generation, especially the male head of household and the aspirations of sons and daughters.

These internal sources of tension do depend, however, upon a range of variables: the household's resources in land, livestock and capital; its involvement in non-agricultural activities; and its social aspirations, including the construction of social networks, the value placed upon education and the ambitions for, or responsibilities assigned to, children. Such variables are dependent upon the organisational form, structure and stage in the lifecycle of the household, and these provide significant dimensions of difference between units. Consequently, the household is very much more than a unit of production and consumption 'subjected to' processes of commoditisation and differentiation. As it displays such heterogeneity of form, and as its internal relations are often marked by conflict and struggle, it would therefore seem apposite to question any unitary notion of 'household'.

No community comprises solely a collection of households exclusively dedicated to agricultural production. In Santa Rosa almost 30 per cent of households do not conform to either 'nuclear' or 'extended' forms. The preponderance of woman-headed and other types of household forms (aged couples, single males) within the land-poorest category (comprising exactly half of all units in this group) attests to the instability of the conventionally viewed household as unit of production and reproduction (Deere 1978). As households which are neither 'nuclear' nor 'extended' are characterised above all by inadequate labour power, which severely hampers their ability to produce agricultural commodities from a limited land base, their reproduction must be secured through alternative economic strategies. While some of the younger and stronger members of such households retain a degree of spatial mobility in search of improved personal livelihood prospects, women with young children and older people encumbered with social and physical constraints that severely limit their mobility may pursue a variety of non-agricultural activities that can generate sufficient to ensure their survival.

WOMEN'S DOMAIN

Given the wide range of remunerated and unremunerated tasks performed by women in Santa Rosa, it appears somewhat arbitrary to divide them between domestic and productive spheres, for boundary definition would seem unnecessarily problematic. Redclift (1985) observes how such dualism, especially that associated with the 'informal sector' debate, constructs categories of labour that are treated to specify the complex relationships through which such work is carried out. She elucidates this point through reference to the way that women can produce both use values and exchange values domestically. This can be illustrated empirically, for many women in Santa Rosa occasionally engage in forms of petty trading that involve exchanges of food goods, for example, potatoes for onions, tomatoes or chilli peppers, of which part are consumed within the home and the remainder exchanged for other goods, such as maize. The maize may then be consumed within the home as food or transformed through women's labour into maize beer which is sold within the community.

Despite the methodological difficulties involved, it is analytically useful to distinguish between the variety of reproductive and productive roles in which women engage, providing both are seen as interrelated. Although reproductive duties are socially determined to include the longer-run tasks of reproducing labour (involving the care, health, education and socialisation of children), and the daily maintenance of the physical conditions of existence within the household, an especially heavy burden is presented by the biological functions of child bearing and early nurturing of infants. Productive roles, in contrast, are taken as those which create some form of income for the individual or the household through the production of either use or exchange values.

BIOLOGICAL REPRODUCTION

For the majority of women between 15 and 50 years of age, pregnancy, birth and lactation are regular features of everyday existence. As Bourque and Warren (1981) observe, having children is very positively valued by both men and women in the Andes, though their commentary suggests that perspectives on the sex of children and optimum family size may differ between the parents. In general, childlessness, whether derived from infertility or the lack of a partner, is a plight viewed with sympathy, for people have no other means of support in their old age. Yet pregnancy and child birth impose a strain on women's nutritional and physiological status, especially where there is little control over fertility and conception may quickly follow weaning of a child.

In Santa Rosa, a survey of births and deaths provided 26 cases of reliable data concerning women's fertility history, that is, women who were at such a stage in their lifecycle that they were unlikely to conceive again. As the ages of these women varied substantially they do not represent a cohort, and the data are inadequate to calculate cohort or period fertility figures. Nevertheless, for the 26 cases, the data reveal a wide range of births per woman; from 3 to 12, with an average of 7.9 births per woman. (The regional figures for 1976 was 8.7 births per woman). However, the total of 205 births is divided exactly between those who were still alive in 1982 (103) and those who have died (102). Of the latter group, and making due allowance for memory recall, 51 were reported to have died between the ages of 1 and 5 years. Of the remainder, 20 (19.6 per cent) had died before reaching the age of 1 after surviving at least 1 week; and 21 (20.6 per cent) had died within 1 week of birth, along with the mother.

Extrapolating these figures into the conventional form of expression for the indicator provides an infant mortality rate of 200 per thousand live births and a child mortality rate of almost 250 per thousand. The extraordinarily high rates of infant and child mortality reflect both the poor sanitary and nutritional conditions in the community, and the effective absence of adequate mother-and-child care. The failure of health agencies to disseminate information on nutrition and advise on simple preventative techniques in rural areas is illustrated.

Child birth in Santa Rosa represents a hazard to the well-being of both mother and child. Despite the existence of the hospital, women do not consider delivering their babies there. Yet the number of women who must seek attention following delivery, as a result of infection, haemorrhaging, and placenta retention suggests that home delivery exacts a heavy toll.[1] Should the baby die within a week of birth, it is quickly blessed and buried in the cemetery. The young child's struggle for life is not supported by medical attention; only the strongest survive this period.

Women breastfeed until the child is between 1 and 2 years of age. It is widely believed that this prevents subsequent pregnancy. Thus, beside spending on average almost 6 years of her life pregnant, a woman spends a further 10 years of her life lactating; for a total of 16 years in biological reproduction. All that is entailed by this continues alongside 'productive activity' (Pearson 1987).

As children become older they are expected to render assistance in the household and from about the age of 5, children move into gender differentiated roles. Though both boys and girls collect water and fuel and watch over animals grazing near the home, only girls begin to participate in the preparation of food, washing of clothes and care of younger siblings. The labour of even young children can replace that of their parents (usually the mother) allowing her to take on additional work or to complete her daily tasks more quickly (Collins 1983). Though the attendance of children at the school, up until the age of 12 or 13, restricts their contribution to the household, responsibilities according to gender must still be performed in the early morning or evening. However, the allocation and burden of such tasks is influenced by the composition and resources of the household, with children of the land-poor families expected to make a more substantial contribution to their keep. It is not surprising that it is from this group that young women seek to escape from unpaid, repressive responsibilities to waged domestic employment in Cochabamba.

THE MATERIAL REPRODUCTION OF THE HOUSEHOLD

Despite the assistance that daughters provide, the woman retains overall control of the daily reproduction of the household and is responsible for one of the most time-consuming tasks: food preparation for domestic consumption. Though the staple dishes are generally simple, food preparation demands women's attendance around the cooking hearth for long periods: peeling potatoes, dislodging maize kernels from the cob and grinding maize and peppers. Yet for many women at critical times of the year, preparation of food is preceded by its procurement when the household's supplies have been exhausted. This responsibility takes women away from the domestic hearth and into an important sphere of interaction with other households. In this sense, the majority of women are not confined to their domestic role, isolated from one another in servicing their families' needs. Women in Santa Rosa have a high level of interaction, although circles of friendship and support tend to be formed between women of similar age, status and residential proximity. As women are responsible for storehouse management, they are also able to exchange small amounts of produce between them. Thus, the basis of women's solidarity arises from their common interests of survival in the face of scarcity and other difficulties:

> These may be of access to money and labour, as well as those, such as illness or eviction, which threaten their capacity for daily survival. Most studies have shown that poor women, as well as women who are independent child-rearers, maintain significant female networks which function as daily or weekly or annual safety nets.
>
> (Whitehead 1984: 6).

Personal observations reveal that prior to the harvest of the early irrigated potato crop, domestic food stocks among many of the land-poor households can fall very low, in which case it is the woman who seeks to replenish the store of potatoes through representations to richer households or to a large landowner nearby. Such credit is generally forthcoming though often on the basis that her partner's labour is available to pay the debt at a later date. For most regular transactions involving cash and exchange of products, women possess a high degree of autonomy over the domestic budget. However, this does vary according to the nature of the 'conjugal contract' within households.

Whitehead uses the term 'conjugal contract' to refer to 'the terms on which husbands and wives exchange goods, incomes, and services, including labour, within the household' (1981: 88). This is a useful concept for it suggests that women and men possess areas of independent economic action within which they are able to make their own decisions with reference to market criteria (Ellis 1988). The product of these individual actions, in the form of cash or use values, is then contributed towards household subsistence. The way in which cash and other resources are created will naturally vary between households, as will the way in which they are divided and the priority of claims upon them.

PRODUCTION OF USE AND EXCHANGE VALUES

It was proposed earlier that women's participation in agricultural production had become marginalised as a result of the intensification of potato cultivation. Beside the performance of tasks associated with field cultivation, however, agricultural work involves the pasturing of livestock. While children are allocated this role if or when they are available, the management of domestic animals falls within women's domain.

Women are generally excluded from dealing with oxen and sometimes horses and mules, but take responsibility for non-working cattle, sheep and fowl. These animals have an extremely low productivity and are rarely sold, although occasionally one may be exchanged. Rather, they serve as a savings bank that can be sold in time of need. The sheep are not systematically sheared for their wool. Most wool is derived from the fleece of slaughtered animals or acquired in exchange with peasants from the high *puna* villages.

The spinning of wool is a ubiquitous activity; women are rarely seen without a drop-spindle dangling from their hands. There are a multitude of activities which women perform that allow them to continue to spin thread simultaneously, though not all women themselves turn their stock of thread into woven cloth. Weaving is one of the first crafts to be undermined by the widespread availability of manufactured goods, and increasingly comes to fulfil a largely symbolic role whose techniques are no longer learnt by the young. However, in contrast to the time-consuming, use-value production of weaving, the brewing of maize beer offers the most lucrative activity under women's control with petty trading a popular, though less remunerative, alternative.

Trading activities perform an important function for many households, particularly the land-poor for whom they can generate a vital source of income at critical times of the year. It is common for women in female-headed households – mothers and daughters – to secure some part of their income from petty trade, the proportion depending upon the role of agriculture and, more specifically, their control of land and access to male labour. The poorest households engage in trading operations involving the smallest capital outlay, while, among the wealthier, commercial trading ventures require more substantial investment. However, while barter and the exchange of goods characterises the trading activities of women, men use cash to purchase livestock either for household use or for resale in the market.

Beer brewing

The procedure for brewing maize beer (*chicha*) is handed down from mother to daughter and remains firmly under women's control. The manufacturing process requires careful skill and judgement and is also physically demanding and tiring. The first day's work, which begins at three or four in the morning, is not complete until after midnight of the following day. During this time up to 500 litres of liquid are moved backwards and forwards between containers. Although the woman retains control of the entire operation, other household members are expected to contribute labour, with older sons involved in the heavier tasks.

Prior to commencing the beer-making operation, a large quantity of wood must be gathered in order to maintain the fire under the large cooking pans (shaped like giant woks) for almost two full days. The fuel wood, comprised of bush for light, rapid heat and heavier logs for long, slow burning, is gathered from a forested area in the higher reaches of the community lands and requires three person-days with horse to cut and carry sufficient wood for the entire process. Those women without sons or male partner in their household, hire labour to provide the fuel wood but, with the help of other kin or neighbours, manage the remainder of the operation themselves, including carrying water from the irrigation ditch to the house.

Chicha is often produced to coincide with religious and secular *fiestas*, but may be produced at other times, especially if a household urgently needs to generate income. However, the transaction between producers and consumers of *chicha* is complicated by the incomplete monetarisation of the local economy and this prevents a straightforward realisation of the potential profit that can be derived from *chicha* production. On the one hand, this allows the land-rich *chicha*-producing households to secure labour power in payment of debts accumulated from drinking sprees, however, on the other hand, land-poor households usually have little need for labour and prefer full payment in cash. Often agricultural produce is used to settle debts. Several producers commented that approximately 25 per cent of potential earnings are lost in unrecovered debts and this would be much higher if the women producers failed to visit the drinkers who owe them money and harass them for payment.

The circumstances which encourage or constrain women's involvement in such activities as *chicha* production, petty trading or even wage labour are complex and wide-ranging. In some cases women's income-generating activities serve to complement the male-dominated agricultural sphere by securing labour through forms of debt-bondage. Elsewhere, such activities provide an important supplementary source of income where the returns from agriculture are insufficient to ensure household reproduction. In other words, there exists a diversity of circumstances in which women make particular choices and develop responsive strategies. In order to illustrate the importance of such considerations within an appreciation of the individual struggles of women, the following section provides two brief case studies.

CASE STUDIES

A single woman

The closest illustration of a subsistence ideal in Santa Rosa is provided by Juliana, a woman in her late forties who lives alone. Here she has about half a hectare of land and some livestock: 15 sheep, a cow, a plough ox, and 4 hens. A second ox has recently died and the number of hens has been reduced by the nocturnal visit of a mountain cat. The land was inherited from her father Dionicio, who was awarded over seven hectares under the Agrarian Reform, but the bulk of his land passed to her brother, Pablo. Pablo died in April 1982 in an accident following a heavy drinking session. Pablo's family live close to Juliana and they regularly exchange labour: Pablo had always preformed the ploughing operations on Juliana's land, and this was now being undertaken by his eldest son, Nicolas. In return, Juliana contributes her labour at planting and harvesting times. She also works occasionally for other households, usually in the 'women's tasks' of sowing potato seed and sorting the harvested tubers under the arrangement of *papa tarpuja*, where she is paid in kind. Juliana is part of a small network of women who share a similar position to herself and with these women, who live in adjacent communities, she freely exchanges her labour. Beside ploughing, from which women are strictly excluded, most of the remaining agricultural tasks on her land are performed by Juliana and her women friends.

During the agricultural year 1981–2, Juliana produced approximately 50 kilos of wheat, about the same of barley and 150 kilos of potatoes which was her share of a harvest divided with her *compañera*. She intended to sow a crop of broad beans, wheat and potatoes – the latter in a share-cropping arrangement – during the subsequent agricultural cycle. Juliana's priority is to meet her own food needs, including exchanging small quantities of her own produce for other goods such as chilli peppers and onions, without resorting to cash. She is able to do this through a social network comprised of close friends and kin, with her brother's family providing an important function in terms of male labour and other forms of support. She is, nevertheless, financially autonomous, with the bulk of her cash income during 1981–82 provided by the one-off sale of a carrying cloth for 650 pesos (US$26) to an outsider, a social promoter who visited the community on several occasions. Juliana spins and weaves the wool from her sheep, which she spends much of her time pasturing. In this way, she clothes herself and, as in this particular year, sells or exchanges a blanket, carrying cloth or potato sack to cover most of her necessary expenses. The only other source of income is derived from occasionally selling eggs to the co-operative store at 2 pesos each, providing about 15–20 pesos per month. Yet, although her cash income is low, so are her expenses, with contributions to the syndicate (100 pesos in 1981) and small expenditures on kerosene, aniline dyes and the occasional glass of *chicha*.

A woman with children

In mid 1981, Dona Julia, a widow of 47 years with three children, was working as a cook for the local landowner, Juan Betancur, for whom her husband had worked as a labourer until his death in 1979. She was remunerated in food and a small cash wage for this work which provided her with the income to support her two youngest children, a boy of 8 and daughter of 3. Meanwhile, Leboria, her 19-year-old daughter, had begun work as a domestic servant in Cochabamba, in the home of a neighbour's sister who had left Santa Rosa many years before. Leboria earned 500 pesos (US$20) per month. This household was reproducing itself entirely from wage labour, an unusual situation in Santa Rosa. Although Aurelio (Julia's husband) had received an allocation of 3 hectares of land under the Reform, some of this had been leased and sold by him by 1979. In 1981, Julia sold one of the remaining parcels of land, measuring three quarters of a hectare for 6,000 pesos

(US$240) and the following year leased another parcel to the same household for 1,000 pesos (US$40). This has left her with one-third of a hectare which remains in fallow.

In September 1981, Julia lost her job as cook as a result of her alleged behaviour during a weekend of drunkenness, decadence and infidelity. It is not clear what triggered such excesses, beside the availability of *chicha*, but a number of people were arrested and sentenced by Betancur, the landowner, who acts as local moral guardian. Beside illustrating the self-appointed authority which local elites take upon themselves, the incident highlights the difficulties facing a woman who transgresses social norms regarding behaviour, for Julia is unable to draw upon the support of other women in the community who regard her with suspicion.

Following the loss of her job as cook, Julia turned to the more popular and reliable combination of income-generating activities: petty trading and *chicha* production. In early October she travelled to Quillacollo and spent 450 pesos (US$18) on lard and onions. On returning to Santa Rosa, Julia spent one week visiting households in the locality exchanging small quantities of these products for maize, acquiring in total some 200 kilos with a market value of 1,800 pesos. She secured the use of utensils for making *chicha* from her next-door neighbour, Amelia. The *chicha* was prepared in time for All Saints Day festival, although this failed to yield the level of profit which she might have hoped for. Though reputedly clearing a profit of 2,000 pesos, it also gave her the opportunity to stay drunk for a week, a danger which most *chicheras* strictly avoid. Julia's efforts to sustain a regular cash income found her developing a new activity: making bread. She purchased a 50-kilo sack of flour in Cochabamba for 600 pesos and baked bread on three separate occasions, selling the rolls in the village and making an overall profit of 750 pesos.

Although the bulk of Julia'a monetary income during 1981–2 was derived from petty production and wage labour she also worked a total of 15 days in agricultural activities receiving payment in kind. She can also attribute her survival and that of her younger children to the proceeds from the sale and lease of land which, although not worked since Aurelio's death, represented one possible source of income through share cropping. It was clear that the remaining third of a hectare was soon likely to follow the route of the other land parcels, passing out of her control in order to meet the needs of day-to-day survival.

Yet, even this was thrown into question when, in mid 1982, Leboria underwent a medical examination in Cochabamba and cancer was diagnosed. An operation was performed, costing 15,000 pesos (US$350), which was met by Leboria's employer under the condition that Julia repaid her within the year. Leboria died several months later, leaving Julia with an insurmountable debt and with few resources that might have allowed her to meet even part of it. Her vulnerability to crisis was now perfectly exemplified as a single woman with two young children and without assets to turn into cash. Both of her children had emerged from the critical phases of early childhood, which had seen the death of seven of Julia's other children, and required expenditure on food, clothing, and education, yet were some way from being economically productive. Her status in the community was low, with limited networks of kin (only a poor and elderly sister survived) or friendship, and no-one appeared able or willing to assist her. People said that this was 'God's punishment' for her behaviour, an expression that probably sought to justify their lack of support for someone in need and a sense of relief that it was not themselves who were suffering in such a way.

CONCLUSION

The case studies illustrate the importance of understanding the dynamics of individual lives if we are to avoid the inappropriate application of functionalist household models. For example, although both households comprise women without a spouse, simply applying the label 'female-headed' is shown to be notably unrevealing. It disguises a considerable diversity of individual circumstances and the range of women's responses in dealing with their situations. Any unitary notion of 'the household' as an autonomous unit has been proved to be dangerously misleading even though, for reasons of brevity, the importance of wider external relations have been under-emphasised. The case studies have attempted to sketch out the types of sociocultural constraints faced by women, within the context of the natural unfolding of the household's lifecycle and deeper processes of structural change.

19

WOMEN'S ROLES IN COLONISATION

A Colombian case study

Donny Meertens

INTRODUCTION

Colonisation makes a deep imprint on family life, affecting the role of labour within the peasant production unit. At the outset men imagine a paradise, goaded on by a spirit of adventure and their visions of future prosperity. Women, instead, express the negative impact of their first encounter with the jungle. They feel disenchanted and betrayed by their new destinies which in general have been chosen exclusively by male family members. Obliged to think of their families' immediate survival needs, women see no paradise at all but rather a hostile environment which is difficult to cope with. Perhaps the most frequent remark heard is: 'everything here was very ugly before because it was all jungle'. The contrast in outlooks between men and women softens and converges with time as the settlement develops.

What are the specific conditions the settlement process generates that lead to such varied interpretations of reality? This chapter follows the migration history of the peasant family in order to explore women's roles during the various episodes in the productive and reproductive work typical of the pioneer family unit. Information was gathered from surveys and testimonies of 77 peasant families living in 16 villages in the township of San Jose del Guaviare (Figure 19.1).[1] These villages are part of a 20-year old colonisation project. The dynamics of the now-distant colonisation front are still represented, however, in the life histories of the protagonists. A limitation exists, inherent in every such research project, in that we addressed ourselves to the successful pioneer families and were not able to consider those who abandoned their endeavour, having failed to settle in the region, and who either returned whence they came or pressed on following the trail of 'primary' colonists. Thus, this study involves a relatively consolidated pioneer population settled in a zone known as 'the colony of El Retorno'.

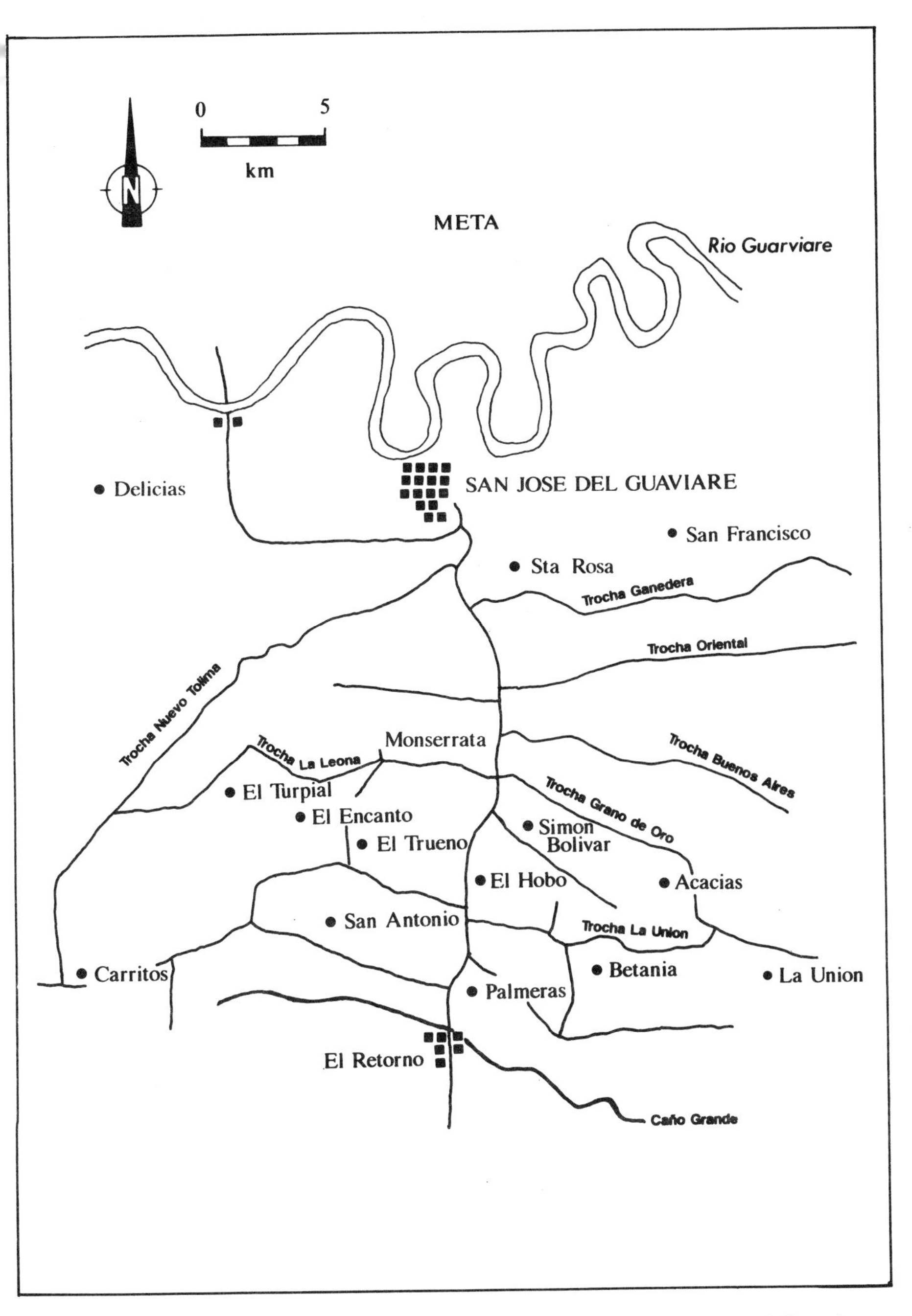

Figure 19.1 Location of villages studied in the region of San Jose Del Guaviare

THE COLONISATION OF EL RETORNO

The Guaviare has experienced several pioneer flows. In the 1930s there was an incursion of adventurers and traders attracted to rubber and they were followed by hunters of pumas. Only in the late 1950s did some latifundias develop in the plains surrounding San Jose. The following decade saw several different settlement streams originating in the agrarian conflict during the civil turmoil known as 'The Violence'. Firstly, there was armed colonisation carried out by peasant self-defence groups oriented to the Communist party. These groups settled on the plains, the Sierra de la Macarena and along the Guayabero River, leaving their mark mainly on the river banks. Back from the river the 'tierra firme' is the site of another colony known as El Retorno settlement which started in 1968. This was promoted by the Vaupes Commissariat and by the journalist Orlando Lopez Garcia, director of a national radio broadcast devoted to the countryside. Far from being institutional, as it has been mistakenly called, it was characterised by the complete improvisation of one individual and the absolute absence of state institutions.[2]

The first step for the settlement was the literal life-or-death struggle of 700 families abandoned in the jungle. The site could only be reached after a very long hike along trails cut through the forest. Some of the settlers were reported to be of urban origin with no agrarian vocation or experience. Many died of starvation or malaria or returned to their former homes or pushed on to unknown destinations (Castro 1976; Santamaria 1978). Others survived, founded homesteads, worked, accumulated belongings and eventually consolidated their struggle. This consolidation was always precarious and often affected by the booms and recessions of the last two decades.

Pastures for livestock were gradually increased. In 1976 an unexpectedly large maize harvest overwhelmed the marketing system and the grain ended up rotting because of lack of transport and adequate storage facilities. This opened the door to a new product, cocaine, which marked a watershed in the history of the Guaviare.

The cultivation of coca plants began in 1978 and increased steadily until 1982 when the first fall in prices occurred as a result of repression and overproduction. Nothing remains for the peasant families from this bonanza which was characterised by the violent presence of narcotics-traffickers, by inordinate expansion in the number of day workers, prostitutes, commercial agents and adventurers and by high consumption of luxury goods. By 1986, when the second coca boom took place, the situation had changed, principally due to the organisational and regulatory (but also violent) presence of the guerillas. Families of settlers process coca on their farms, sell this cocaine base and reinvest the earnings in cattle. Now, a new depression in coca has set in and there seems little prospect of recovery.

Of the families in the sample, the majority (73 per cent) had arrived before 1974, that is, during the early development of El Retorno from 1968 to 1970 or the period immediately following, 1971–3, when the opening of the road between Puerto Lleras in Meta and San Jose del Guaviare gave a further impulse to the migratory flow. Settlers differed in origin and migration incentives. The first group, 16 per cent of the sample, had little in common except their poverty and the fact that they had been recruited through Lopez Garcia's radio programme. The second group, arriving between 1969 and 1970, came from Boyaca and were steeped in Conservative party tradition. They had been encouraged to migrate by one of their political leaders in order to reinforce their party's political presence on the colonisation frontier. The third group is composed of those informed or invited by relatives or acquaintances already established in the settlement. They reinforce the characteristics of the initial settlers since their origin is similar.

ORIGINS OF SETTLER FAMILIES AND THE ROLE OF PEASANT WOMEN

It is impossible to analyse the impact of colonisations on peasant women without first knowing the characteristics of their places of origin. One-third of the men and women settlers came from Boyaca, with the next most popular migrant sources being the provinces of Santander, Meta and Cundinamarca and finally the coffee zones of Viejo Caldas and Tolima (Table 19.1, Figure 19.2). Only one respondent, a woman, came

Table 19.1 Place of origin of settler couples in the Guaviare

	Women		*Men*	
Place of origin	*Number*	*%*	*Number*	*%*
Boyaca	23	29.9	26	33.8
Santander	17	22.1	15	19.5
Meta	10	13.0	5	6.5
Cundinamarca	8	10.4	10	13.0
Viejo Caldas	6	7.8	10	13.0
Tolima	4	5.2	3	3.9
Other rural area	8	10.4	5	6.5
Urban	1	1.3	–	–
Total	77[a]	100.0	76	100.0[b]

Source: Author's fieldwork
Note: [a] In one case there was no husband or partner as the respondent was a widow who had come to the Guaviare alone and did not remarry
[b] Some totals are rounded

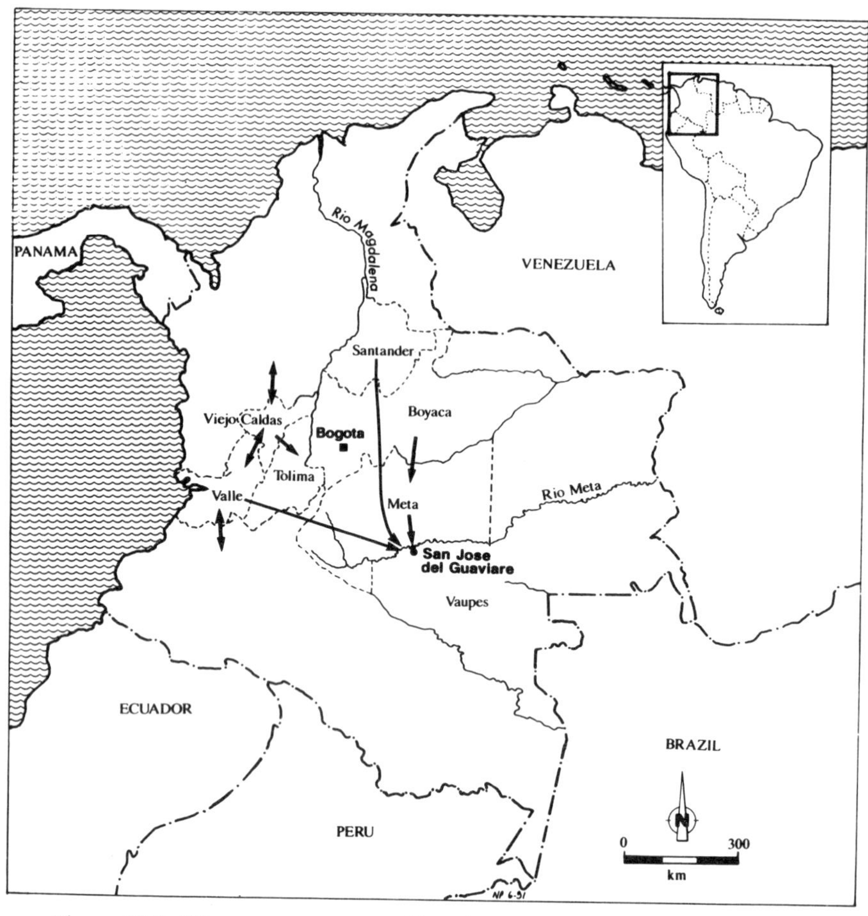

Figure 19.2 Migratory patterns of immigrants to San Jose Del Guaviare

from an urban area, contrary to previous reports. Successful colonisation appears to require both men and women working together, but one interviewee, a widow, had migrated without a partner and remained single. Her presence reinforces this study's findings concerning the important role played by women on the pioneer frontier.

The predominance of settlers from Boyaca may be explained by the landownership structure of the province and the progressive deterioration of its soils which have given Boyaca, together with Cundinamarca, the highest level of outmigration in Colombia. We found that most of the communities of origin were located at an altitude of between 1,000 and 2,000 metres where sugar cane, plantain, cassava, maize and some coffee are usually grown and where livestock is important. This finding dispels the widely propagated notion that the settlers were totally unadapted to the new environment because they had moved from the high, cold altiplano. A quarter of the settlers reported having had some land but not enough to support a family and so the main reason for joining the colony was the quest for land.

The work of peasant women has always been important in temperate Andean regions where there is semi-proletarianisation because of the sub-family-sized farms. Before migrating the women had been day labourers, tenants or managers of small farms. Women have traditionally been involved in farm work, especially in tasks requiring fine manual skills and allowing work in short spans such as sowing and harvesting. Where men work off the farm, women will carry out all farm activities. Cattle in many parts of the Andes are cared for by women in contrast to the pattern on the farms of the lowlands (see Townsend, Chapter 20). On small plots the livestock may be exclusively entrusted to women. On medium-sized farms women share in cattle raising, tending even to withdraw from agrarian chores in order to devote themselves to livestock, specifically to milking, feeding and surveillance. This zonal contrast is marked not only for Colombia but also for Peru, Bolivia and Brazil.[3] Explanations may be linked to the more aggressive behaviour of zebu cattle in warmer climates and the introduction of capitalist relations of production on lowland ranches requiring male wage labourers (cowboys). The biological flexibility inherent in livestock and the combination of its high use value and high exchange value make cattle raising on a small scale fit easily into the peasant economy where animal care can be combined with other domestic and agrarian chores. This creates the conditions under which livestock management becomes a major female occupation on smallholdings. Thus, women who came to the Guaviare came as peasants, with a tradition of involvement in agriculture, especially livestock raising.

Not all peasant families migrated directly to the Guaviare (Table 19.2). Some had travelled to other regions as seasonal labourers or even to other countries such as Venezuela or Panama. Marked gender differences can be observed in these migratory patterns: more women than men arrived directly in the Guaviere and men had usually travelled greater distances than women. A significant number of women and men (15 per cent and 12 per cent, respectively) had lived in a city, usually Bogota, during their travels. For women this urban stage had almost always been undertaken as an adolescent domestic servant. A site of special convergence of migratory flows focused on the towns of Guamal and Acacias in Meta, a region of early colonisation from Boyaca and Santander where agriculture dominates in contrast to the cattle ranches elsewhere in the Llanos. In many cases the peasant women had migrated to the Guamal region as children with their parents in search of land.

Table 19.2 Migration itinerary of men and women settlers

Itinerary	*Women* *Number*	%	*Men* *Number*	%
Went directly to Guaviare	29	38	12	16
Migrants with intermediate stopovers:				
Lived in the city	6	8	7	9
Lived in city and Meta	5	7	2	3
Lived in Meta and rural	21	27	39	51
Lived only in rural	16	21	16	21
Total	77	100[a]	76	100

Source: Author's fieldwork
Note: [a] Some totals are rounded

The land colonised was insufficient for the second generation and so the women had to work as day labourers or sharecroppers. Guamal became a point of matrimonial union with a quarter of all the women surveyed joining their husband or partner in this town.

GENDER ROLES AT DIFFERENT STAGES OF COLONISATION

In the process of colonisation two types of colonists may be distinguished. The first is represented by settler families arriving in a particular village and remaining there. Usually they were part of the earliest group of settlers and were able to obtain large allotments of land or they arrived with the capital to buy sufficient land for their needs. The second group have to undergo a step-by-step process of savings and accumulation before being able to acquire a farm and settle down. In general, a family which exhibits spatial mobility within a settlement area has usually arrived with few resources. The survival strategies of these families follow similar patterns. They lodge at first with a relative or friend and do wage work, receiving payments in uncleared land. This first stage lasts about 18 months. The second stage, when the family moves beyond mere subsistence, lasts 4 to 6 years.

But from 1978 onwards, after coca was introduced to the region, once a family acquired a piece of land this second stage became shorter. According to some authors (Hecht 1986; Molano 1987) it is precisely at this juncture that the Guaviare stands alone compared to other more standard settlements in Latin America. Coca production and the presence of guerilla forces counterweigh differentiation and prevent it reaching the extremes of the latifundia/minifundia dichotomy so prevalent in the Andean zone. Around the nucleus of El Retorno, two settlement fronts have probably felt the accumulation and differentiation processes most acutely: the very accessible zone near San Jose and the central road, and the remote dynamic frontier thriving wholly on cocaine.

Farms in the survey averaged 32 hectares (80 acres) with the range being from 0.4 to 160 hectares. All families surveyed had some document proving landownership and 62 per cent held a title deed from INCORA, the national land reform agency (Table 19.3). Only 15 women held land titles, which is considerably fewer than is normal in the Andean region. Of the women holding titles, 80 per cent (12) had acquired them through inheritance, generally after the violent deaths of their husbands. The married women who owned no land excused themselves on the basis of illiteracy or because they lacked identity papers or a decision-making role in the family.

Table 19.3 Landownership according to gender

Title holder	*Number*	%
Men	47	61.0
Women	13	16.9
Both	6+2[a]	10.4
Children	3	3.9
No formal deeds	6	7.8
Total	77	100.0

Source: Author's fieldwork
Note: [a] In two cases both partners had their own land and title deed

As farms become larger over time, women's labour input decreases in intensity as economic conditions permit the hiring of wage labourers. Women settlers on larger farms also devote themselves more to livestock than to agriculture as tending cattle is felt to be a more suitable occupation. Female involvement in agriculture remains high where share cropping has evolved.

At the early stage of colonisation, two factors made women's lives hard; the first was the type of work demanded and the second was the stage in the lifecycle. The opening up of new land and the precarious nature of survival until the first crop was harvested required everybody's occupation.

> When we staked out this land the whole family had to cut down the mountain, clear out the timber and the weeds, plant rice, corn and plantain, then help harvest, lay out pasture, build fences and raise cattle. It didn't matter if we were women, Sundays were destined for laundry, cutting firewood, shelling rice and getting food. We spent around 5 years doing it.
>
> (Woman settler)

Clearly, during this stage women broadened the scope of their participation in agriculture even beyond the already high levels of involvement most of them had as a legacy from their places of origin. With regard to the family lifecycle, it is generally found that families opt for migration when the children are very small or even during the first pregnancy. Interviews clearly linked the decision to settle on new land to the arrival of the first child, pinpointing this as the moment of crisis in terms of resources.[4] Most women were 13 to 20 years old, the age of family formation, when they arrived in Guaviare. 'Work before was very hard, especially for the newcomers, because the children were very young and farm work is very heavy. One had to clear and burn land and at the same time to cook, tend to the workers and raise the children.'

At the second stage of settlement, not only does women's work intensity slacken as the rhythm of land clearance eases, but the children grow older and are able to help. Furthermore, coca production has brought greater financial solvency and allowed extra hands to be hired. Workers are hired principally to harvest coca leaves. In rare cases a cook may be hired to make food for the farm workers, an important and time-consuming job normally carried out by the housewife. Work is considered to be lighter in the Guaviare than in the Andean region in two respects: in agriculture because less weeding is necessary, and in livestock because animals do not need to be tethered or to be under constant surveillance.

The fact that women work less now than in the early stages of the colony does not mean that they have lost status or been progressively marginalised from agricultural and livestock production.[5] On the contrary, the initial broadening of the extent of women's involvement in production has been maintained. Even if women do not work in these jobs permanently they have the experience, capacity and social acceptance to take them on in an emergency, replacing their husbands when necessary. Peasant women insist that their work has more recognition from the men now than before the settlement process. Survival in a colony is so precarious nobody can afford to be a non-participant and men feel the need of a woman's support for success. The greater recognition of the crucial importance of woman's work in the survival of the family unit has not been accompanied by a concomitant broadening of male participation in domestic activities.

THE CONTEMPORARY GENDER DIVISION OF LABOUR IN FARM WORK

Peasant women carry out a great diversity of tasks during the day, some more visible than others, all of them frequently interrupted by another activity. One working day is not 'just one job' for women but a permanent combination of tasks and allotment of attention and energy. This diversity requires an interpretation of her participation not only in

Table 19.4 The gender division of labour in Guaviare

Domain	*Women only*	*Men only*	*Both*
Domestic labour	90%[a]–housekeeping –childcare –cooking for workers	0%	10%–childcare
Vegetable garden	35%–seedbeds –sowing –transplanting –weeding, cleaning –fumigating –irrigating –harvesting	0%	65%–fencing –ridging –furrowing
Small animal husbandry	76%–feeding –medication –nesting –collecting eggs	0%	24%–feeding –medication
Traditional crops	9%–shelling maize –seedbeds(cocoa) –transport(cocoa) –drying(cocoa) –podding(cocoa)	9%–pasture upkeep –fumigating –fertilising	82%–sowing –weeding –harvesting –burning –cutting –worker supervision –sugar milling
Coca	13%–cooking	9%–laboratory	78%–harvesting
Livestock	13%–milking –surveillance –health care	12%–pastures –fencing	75%–cleaning –worming –branding –castrating –herding –vaccinating
Marketing	10%–coca –eggs –dairy produce	18%–cattle	72%–loafsugar –cassava,rice –plantain, –fruit,cocoa –maize
Transport	13%	39%	48%
Communal action	18%–committees –parent's assoc. –action groups	28%–same	54%–same

Source: Author's fieldwork
Note: [a] Percentages refer to the number of cases recorded in each mode, N=77

the traditional terms of 'productive' and 'reproductive' work but also discriminating information by work areas, crops and activities on the farm (Deere and Leon 1982). Table 19.4 describes these activities.

Many tasks are reserved according to gender. In other cases the division of labour is described in terms of excluding women from certain activities. Traditionally, for example, women should not use a hoe on the ground, till, open trenches, pack mules or cut timber. In other words, women should not open new lands or initiate agricultural cycles. In the settlement process, as we have seen, these limiting norms were exceeded and even if subsequently, the toughness of women's work has diminished, it does not mean that women have returned to traditional labour patterns. Evidently, as the different testimonies show, there is more recognition now of women's invisible work, although there still persists an underestimation of the productive character of many of their activities.

Women continue to work longer shifts than men: on average women work 16 hours a day and men, 14 hours. In extreme cases women work 4 or 5 hours longer than men, especially where a widow with a large holding is concerned, or when a woman works for wages in the fields and additionally must perform household chores and tend to the ever-present domestic animals, thus completing the triple load.

One of the activities that most determine women's work is cooking for wage labourers, who are hired in large numbers every six weeks, to harvest coca leaves. On average, a minimum of 6 and a maximum of 22 wage workers can be found on a farm. Preparing food for these workers is an activity exclusively carried out by the woman of the house and one in which men never take part. She can count on the aid of the children to carry out collateral activities such as carrying water and taking food up to the fields. But it continues to keep a woman from participating in tasks or events that imply her being absent from the farm, unless her daughters are old enough or there is enough income to hire a cook to replace her. Adhering to the conventional definition of productive and non-productive activities, a woman who remains in the kitchen would be replaced in the latter category and she would be called a 'housewife'. But since she not only does the cooking for her own family but also for the workers, she is, in practice, contributing with her work to a share of their salaries which implies savings for the family unit. For this reason, this is one example of how female work can, even when domestic, have a productive component that carries with it implications for the balance of income and expenses on a farm; for the rationalisation of its costs of operation and, in cases of precarious peasant economies, for the very survival of the unit.

Domestic work confined strictly to reproduction, is also an exclusive female domain. Only in 10 per cent of the couples did men participate in tasks related to the children's health or recreation. This is quite different from the situation found by Townsend (Townsend and Wilson de Acosta 1987) in the San Lucus Serrania, where the demographic structure resulting from the migration of young women to domestic service in the city, has forced men to share in the domestic work at home (see Chapter 20). In the Guaviare, 42 per cent of families with children over 12 years had members who had migrated for study or work purposes but there was no gender bias in the migration flow. The population pyramid for settler families shows no great gender inequalities so, in this case, we can discard demography as a decisive factor in the gender division of labour.

Agricultural activities taking place around the house, such as the cultivation of vegetable gardens and orchards, and animal care, are mainly considered female responsibilities. Men help with some specific chores. In gardening, they do the fencing, ridging and furrowing. Domestic animals receive only minimal attention on the rare occasions when men supply food or medicine. As far as fruit orchards are concerned, there is no rigid division of labour since both men and women participate throughout the production cycle.

The cultivation of traditional crops, which except for cocoa are seldom commercialised, is a fundamentally shared workload. Most (91 per cent) women take part in these activities and in the rare instances when they do not, they dedicate themselves to livestock raising. Both men and women work with maize, plantain, cassava, sugar cane and cocoa, whereas pasture and rice fields are preferentially male-dominated areas.

Instead, women are responsible for crops of lesser importance, in terms of quantity and commercialisation, but which, nonetheless, fulfil a role in the production system, be it by complementing the family diet (like pumpkin, beans, pineapple or coffee) or as animal fodder. Women also participate in the production and processing of coca, sometimes harvesting the leaves, but particularly in the specific phase of cooking leaves, while men are in charge of preparing the chemical mixtures to be added.

Women are involved in the processing of other harvested produce. Two specific phases of cocoa production are in female hands: at the beginning, making seedbeds and transplanting seedlings, and after harvesting, podding the cocoa beans and drying them in the sun. Males predominate in land clearing. But the imprint of the early stages of settlement remains and in all stages of agrarian and livestock production there is no real gender exclusivity.

Andean traditions of cattle raising are reflected in high female involvement on the frontier of colonisation, with 88 per cent of women in the sample working with livestock. Women often withdrew from other agricultural activities when the farm was well established and the children old enough but they did not want to give up stock raising. Although traditionally they carry out such activities as milking, surveillance, folding and health care, the women of Guaviare also undertake tasks considered masculine, such as breeding, cleaning and vaccinating.

Marketing farm products is carried out by men and women separately. They devote themselves to the selling of different products.

Buying and selling cattle and wood are considered male domains, while women sell eggs, dairy products and domestic animals. Women prepare goods for sale by shelling, selecting, packaging and sometimes also direct retailing. But their greatest contribution is to cocaine commercialisation and distribution where they act as 'mules'. Other products are sold by women at the farmgate or locally.

The division of labour in the study area is similar to that in the Andean zone in three ways. The restriction of female participation to activities related to self-sufficiency, except for coca paste, is a repetition of the division of labour in the Andean zone where women produce use values and men produce exchange values. Women in Guaviare also carry out tasks that can be done around the house, that can be interrupted easily and that require fine manual skills. The third similarity refers to the exclusion of women from certain activities, such as the upkeep of pasture land.

The division of labour in Guaviare also differs in three ways from that of the areas of origin of the migrants. Firstly, and most generally, is the lack of rigidity in the daily pattern of the gender division of labour. The second difference can be seen in a broadening of the norms regarding female participation in rough and supposedly virile activities, such as the clearing of land. The third distinctive characteristic is the sustained participation of women in cattle raising, which even though continuing the family-cow tradition of the former peasant economy, defies the living habits of warmer climates where most cattle ranches and the more temperamental breeds are located.

Finally, we must consider the last item in Table 19.4, women's involvement in community action. Peasant women in Guaviare do play a key role in agriculture and livestock production but the same does not hold true for community action. Although many women do participate, singly or with husbands or sons, one-quarter of the women surveyed always delegated involvement in the public sphere to the male members of their families, justifying their passivity with expressions of shyness and low self-esteem, that contrasts sharply with their conscious role in productive farm life. But other women are deeply involved, with many of them holding leadership positions on the committees. Perhaps this dichotomy reflects both factors of the settlement process: on the one hand the preservation of traditional values, inherited from their places of origin and continued in those spheres of life in which pressure for change has not been strong or pervasive enough (like community action); and on the other hand, women who have maximised the dynamics engendered by the settlement process and the broadening of their productive roles, in order to also break with former inhibitions, responding positively to the different stimuli of institutional, political, and cultural community action. Just as the pressing need to survive pushed the traditional barriers of the division of labour forward, so too can supporting women's organisations, training and enrolment in extension programmes change traditional patterns of gender differences in public interaction in the community sphere.

CONCLUSION

We must now examine the extent to which the changes in the gender division of labour brought about through the impact of colonisation have become generalised across all families. The geographical origins of the migrants in temperate latitudes of the Andean region as peasants on tiny holdings or semi-proletarian sharecroppers, provide guidelines on the role of these women before migration. According to available literature these source regions are characterised by a very high level of female involvement in agricultural activities and exceptionally outstanding female participation in cattle raising. This cultural heritage in the definition of male and female roles has facilitated peasant women's insertion in the exacting task of colonising.

Colonisation has also had its own impact on women's participation, broadening their scope of action in activities relating to the opening up of new land and the initiation of agricultural cycles. The first settlement stage not only signalled female involvement in new activities but also an exceptional juncture of factors which meant a period of extreme hardship and labour intensity for women. With time, the intensity of women's work diminished as children grew older and were able to help, land clearance occurred less frequently and production of coca provided the capital to allow for the hiring of workers to replace family labour. However, women have not been marginalised from productive tasks and are now able to replace their husbands in any area of production or management of the farm whenever necessary. Yet role definitions remain rigid when it comes to men participating in areas traditionally considered to be female.

In spite of this one-sided rigidity in the division of labour, a significant aspect of the colonisation experience has been the increased flexibility of women's roles. The results of this study differ from those of other colonisation zones in Latin America, perhaps because of the farming systems in the areas of origin of the migrants or the importance of coca growing in the Guaviare settlements. It therefore becomes imperative to take advantage of the broader role definition noted in this study in order to encourage more autonomous female participation in regional development.

20

HOUSEWIFISATION AND COLONISATION IN THE COLOMBIAN RAINFOREST

Janet Townsend

INTRODUCTION

Housewifisation (Mies 1986) has characterised three populations of colonists I have surveyed in Colombia. That is, with migration into areas on the frontier of settlement, women have lost much of their access to economic activity and have become housewives on farms, engaged full-time and overtime in social and biological reproduction. In Latin America, there is evidence that this has occurred in planned and spontaneous rural settlements; elsewhere, it has been reported for planned settlement schemes in Nigeria (Burfisher and Horenstein 1985), Kenya (Hanger and Moris 1967), Zimbabwe (Jacobs 1984) and Sri Lanka (Ulluwishewa 1989).

In Latin America, four patterns of gender roles in colonisation appear in the literature, with different processes dominating change.

1 *The farming system*: Judith Lisansky (1979) found that in the Mato Grosso (Brazil), gender roles change dramatically over time as the farming system evolves. The first settlers create family farms in which women contribute significantly to production; the farms are homesteads, relying mainly on crop production. Ranches then displace the farms, and women lose their income-generating roles in the countryside and are reduced to housewives and prostitutes. This pattern is also reported for ranching in Colombia by Clara Gonzalez (1980) and in Amazonia by Susanna Hecht (1985). In this pattern, it is capitalist penetration and the transition to ranching which lead to housewifisation.
2 *The life cycle*: In the Guaviare (eastern Colombia), Donny Meertens (1988, 1989) found a very different pattern (see chapter 19)
3 *Initial migration*: In Santa Cruz (Eastern Bolivia), Susan Hamilton (1986) found that women in the highlands have extensive income-generating opportunities but that they become housewives when they

migrate to the lowlands. The change comes at the moment of migration, not in subsequent development in the new area. She suggests that in migration, women 'may lose control over former sources of responsibility and independence, such as flocks of animals or yearly stores of crops'. (The move into a cash economy and a different culture is also important.) This interpretation matches accounts of Santa Cruz by Lesley Gill (1987) and James Weil (1980).

4 *Continuing migration*: In the middle Magdalena valley (Colombia), I found that women become housewives when they migrate, not on the arrival of ranching, as in Amazonia. The confinement of women to reproductive roles, however, is greatly compounded by the outmigration of daughters to the cities and the highly masculine sex ratio which results (Townsend and Wilson de Acosta 1987; Townsend 1988).

Despite the very extensive colonisation literature, it is difficult to establish how widespread these patterns are. There could be as many constructions of gender as there are colonisation zones. My concern in this chapter is to seek a deeper understanding of the patterns and processes by which colonist women become housewives under very different farming systems in the middle Magdelena valley.

GENDER

Nanneke Redclift (1985) has demonstrated that the gender allocation of tasks in production and reproduction is neither independent, determined nor determinate in any simple mechanical way. By definition, gender roles in production and reproduction are a dimension of changing *spatial divisions of labour*: local, regional, national and international. Gender in a locality can only be understood in terms of past and present insertion of that locality into spatial divisions of labour. Doreen Massey (1984) has described the structure of a local economy and society, very schematically, as the product of a combination of layers, each layer representing the successive imposition over the years of new forms of activity. In a capitalist world economy, each layer also represents a round of investment (or disinvestment) by capital within a new spatial structure of production. In a given locality, the layers together represent the succession of roles which the local economy has played within wider national and international spatial structures. New spatial divisions of labour will be determined in part by the existing structure. 'The causes of the contemporary pattern then cannot be sought simply in the current balance of power between different social forces, but must include the sedimented forms of the previous rounds' (Walby 1986).

Many conflicts over interpretations of gender arise from a failure to distinguish between strategic and practical gender needs (Moser 1987). Women, for instance, may have a perceived need for clean water; in fact the whole population has the need, but it becomes a women's need because of the role ascribed to them by society. Women's 'need' for water is a practical need, called for by existing constructions of gender; to enable them to meet the need is not to change society and is not a feminist action. To meet strategic gender needs would be to change society and move to new constructions of gender. It is unreasonable to criticise practical action because it does not change society, or strategic action because it does not meet the immediate requirements of present society; the question is, what are the objectives?

CASE STUDY: MIDDLE MAGDALENA

I have examined the significance of gender in participation in productive and reproductive activities in three surveys in Colombia. In Colombia, the national sex ratio is feminine (98 males per 100 females: 1985 census) but most rural areas are masculine (108 overall). Gender roles, gender relations and migration interact to produce this. The three survey areas differ in their insertion into regional and national structures of production. They include a low-input, high-input and co-operative case. All are in the middle Magdalena valley of Colombia, a region of humid tropical forest which has been largely cleared over the last thirty years.

San Lucas, the low-input case, is a locality where colonisation has stagnated (Townsend and Wilson de Acosta 1987), access to markets is poor, shifting cultivation is still significant and the displacement of small farms by ranches is incomplete. There is differentiation and wage labour, but few families are landless. Human and animal power dominate production and transport. San Lucus is primarily a labour reserve and many girls leave at an early age. Households are large, and so are completed families. The sex ratio among adults (15 years and older) in the 75 households sampled in San Lucas is 189 men:100 women; from the age of 10, females are in a minority.

The other two survey areas are not of 'the frontier' but of 'second-generation' colonisation: they involve settlement schemes in ranching areas. The first, El Distrito, is strongly incorporated into the urban–industrial economy. This is a land reform project, where the state purchased ranches, invested in drainage and divided the land into 211 farms of some 50 hectares. In the 1980s, a World Bank loan was used to develop irrigation. The farmers are engaged both in ranching and in the mechanised production of rain-fed and irrigated rice, with aerial spraying and even some combine harvesters. A third of the families surveyed had their own tractor. Production is energy intensive; there is little place for permanent family labour, and completed families are small. The adult sex ratio in the 30 households sampled is 153 men:100 women.

The other area, La Payoa, is a new project, inspired by a Catholic non-government organisation. A co-operative has been formed by 58 landless families, all from the nearby mountains where sisal production is in crisis. The co-operative has bought 3,000 hectares for US$500,000, and has now to pay the debt. Their aim is to achieve peasant-based production using appropriate technology and with maximum autonomy from the market, but their freedom to withdraw from the market is constrained by the high level of debt. The move began in June 1986, so that the colonists are all new to the area and a new construction of gender may be observed. Teenage girls are mainly planning to leave but older male kin are brought in on a temporary and semi-permanent basis from the home area as extra family labour. Less than 18 months from the inception of the project, the adult sex ratio in the 46 households interviewed was 136 men:100 women.

Class structure differs between the survey areas. If we make a crude division between landowners, merchants, rich peasants hiring labour, middle peasants able to achieve a livelihood with family land and labour, and poor peasants selling labour, then the San Lucas survey has all classes of households, the El Distrito survey only rich peasants, and La Payoa only middle peasants.

HOUSEWIFISATION

The three survey areas contrast sharply in farming systems, land tenure, class structure and market relations. Yet in all cases, gender roles are constructed in such a way as to expel women to the cities and exclude them from income-generation on the farms. What in this context is so conducive to 'housewifisation'?

Gender roles in income generation in these areas almost exclude women (Table 20.1). Activities which add income to the household are overwhelmingly male. In all three survey areas, cattle are important, and colonists maintain that these lowland, zebu cattle are wild, dangerous and unmanageable for women. Highland cattle are seen as more placid and in the highland source areas for all La Payoa colonists, most of those at El Distrito and many in San Lucas, women share in milking and caring for cattle. Source areas are of some significance: in La Payoa, women still share almost equally with the milking (though usually as subordinates) and contribute to work with cattle (Table 20.1).

This disqualification of women from productive roles follows the pattern found by Susanna Hecht (1985). Hecht, however, expected dairying and small animal production to give women a role. Judith

Table 20.1 Ratio of gender participation in selected tasks (women:men; 13 years of age and older)

Task	*San Lucas*	*El Distrito*	*Payoa*
Cook for labourers	43:0	7:1	16:1
Household laundry	10:1	8:1	18:1
Domestic cooking	6:1	8:1	18:1
Feed hens	3:1	8:1	11:1
Fetch water	1:1	4:1	3:1
Feed pigs	1:1	7:1	15:1
Fetch wood	1:6	4:1	3:1
Gardening	1:6	1:3	1:2
Milking	1:2	1:7	1:1.5
Agriculture	1:60	1:52	1:70
Cattle work	0:65	1:28	1:13

Source: Author's fieldwork

Lisansky (1979) found women's participation in production to be considerable at the early colonisation stage on small, crop-producing farms, but even in San Lucas where these occur, this was not found to be so.

The women of La Payoa feel that the move has deprived them of access to income. The colonists were previously landless labourers and sharecroppers in the sisal area. None of the women have ever been substantial earners, but they had paid 'for the children's clothing' and often for food from long hours worked in sisal handicrafts and an array of minor activities. They often had a kitchen garden (only three have one now) and could contribute significantly to nutrition by rearing pigs and poultry (few can still do so). They feel that there is now nothing they can do to increase family income, although they hope and plan to produce pigs and sheep for sale in the city.

Women's main direct contribution to household income in these three areas is, as Carmen Diana Deere and Magdalena Leon found in Colombia and Peru (1983), cooking for labourers (see also Donny Meertens, Chapter 19). The daily farm workers' wage is quoted as a cash amount, or as cash plus three meals, cooked by the women of the farm. The full cash wage is rarely paid for labour, and there are seasons when women rise at 2 a.m. to start cooking (Table 20.1). This certainly increases net household cash income, but does not give women improved access to cash.

Gender roles in reproduction give males more of a role than is common in rural Colombia. There is a gradation (not necessarily causal) from the area with the most masculine population, San Lucas, where reproductive tasks are held in high regard, to the least masculine, La Payoa, where these tasks are more exclusively performed by women. Child care, cooking and laundry account for most of women's time.

Child malnutrition and diarrhoea are widespread in all three areas, but much more severe in San Lucas where children weighing only half of the expected weight for age were encountered. Women's reproductive roles are exceedingly onerous, as housing conditions illustrate (Table 20.2).

Table 20.2 Housing conditions (% of individuals)

	San Lucas			
	Papayal (village)	*Serania (forest)*	*El Distrito (land reform)*	*La Payoa (co-op.)*
House				
Cob	70	18	3	0
Pole	8	23	0	2
Plank	22	57	11	5
Brick	0	0	23	66
Block	0	2	63	27
Stove				
Floor	27	0	0	2
Raised	60	100	24	98
Paraffin	13	0	0	0
Electric	0	0	6	0
Gas	0	0	70	0
Floor				
Earth	54	86	11	18
Cement	46	14	81	82
Tile	0	0	8	0
Roof				
Thatch	70	88	0	3
Wood	0	3	0	0
Tile	0	0	0	20
Tin	30	9	84	45
Asbestos	0	0	16	32
Light source				
Candle	10	5	0	0
Paraffin	37	92	57	100
Electric	53	3	43	0
Water				
Untreated	20	81	55	39
Chemicals	67	9	0	0
Boiled	13	10	45	61
Sanitation				
None	67	95	29	90
Latrine	15	3	0	7
Flush bowl	18	0	0	0
WC	0	2	71	3

Source: Author's fieldwork

GENDER RELATIONS

These surveys have concentrated on gender roles. This is not to accord them analytical priority, but the circumstances of colonisation do pose problems for the study of gender relations. Some lineaments of gender relations are clear. Land tenure is dominantly male, and in all three cases, even La Payoa, when asked about the animals which are the family's main source of livelihood, women may say 'I don't know how many head of cattle *he* may have.'

Isolation is a distinctive feature of women's lives in colonisation. Women's exclusion from agricultural field tasks has more significance than merely excluding them from production. Female activity space is very different from male, since women and girls leave the house mainly to do the laundry. Shopping may provide outings, but in San Lucas, shopping is a male activity in farm families. Women colonists may have little social contact outside the household; in El Distrito, many have never met their women neighbours. This is 'housewifisation' to an exceptional degree. In their home areas, kin had usually been an important source of social contact. As colonists, men leave the farm frequently, to trade, to engage in wage labour, to develop links with neighbours, to engage in social drinking. Many women are without these outlets. Some respondents in San Lucas have delivered their own babies because they 'had no-one to ask for help'. Many speak highly of the satisfactions of seeing no-one but family, but most found the loss of their mother as a neighbour a very negative part of colonisation.

The explanation offered locally is women's family role. Women must stay in the house to care for children. The escape from this isolation, encountered among rich peasants and landowners, lies in the practical requirements of education. La Payoa and El Distrito have schools which cover the first part of primary schooling; San Lucas has some primary facilities, but not enough. For the rest, children must do without, or move to live near a school during term. More prosperous families maintain a village, town or even city dwelling for this purpose, and mother lives there with the school children, usually during term, sometimes permanently. Sometimes the whole family leaves; the original land reform plans of El Distrito required farmers to live on their farms, but with increasing prosperity, the demand for secondary education has led to a relaxation of this rule and allowed absentee ownership.

Male violence is not easy to estimate. In general, colonist men drink, but women do not. There is domestic violence, including killing, but the amount and significance are difficult to evaluate. Alfredo Molano (pers. comm.) has suggested that women's position in colonisation areas is unusually strong. Their presence is critical to the survival of the enterprise and their bargaining position unusually good. Certainly, women's role in reproduction is highly regarded; men and boys engage in reproductive tasks to a very exceptional degree, and speak in unusual terms of their importance. All this, again, is difficult to evaluate and calls for further research into the patriarchal bargains of colonisation. What options are open to women? Is it a simple choice between migration to the city to work in domestic service and a submissive, isolated life? What manipulation and negotiation are open to women who stay?

What are women's practical gender needs in these areas? In La Payoa, which has embryonic women's groups, their needs are vigorously expressed: clean water, preferably piped to the house, latrines, electric light, a clinic (with contraceptive advice), more schooling for children, courses on nutrition and health care and training in income-generating activities (they suggest handicrafts, farmyard animals, kitchen gardens, but not dairying, care of cattle or the keeping of farm accounts). In El Distrito, where physical infrastructure is good, the focus is more on access to facilities of city standard. In San Lucas, medical care and adequate schooling are undreamed of; women's concern is primarily with achieving minimum nutrition for the family. Felt personal needs are very rarely expressed, save by adolescent girls.

To an outsider, it may seem that women's strategic gender needs are for solidarity and for access to income. The women themselves may be more conscious of the positive options offered by the present patriarchal bargains. These seem to exclude women from control over income or even being able to add directly to household income. Women see a positive side to this: they speak highly of the social isolation, and of working always in the shelter of the home. They are glad not to be a labour reserve for productive activities. It may also prove that they have considerable autonomy in the reproductive sphere and even power in family decisions.

CONCLUSION

Unsurprisingly, it appears that the lineaments of gender in colonisation vary greatly between places. Our information remains fragmentary, which is a serious restriction on policy. Quite simply, we know little of women's roles or of gender relations in these areas, and still less of women's practical and strategic gender needs. There are interventions which are obviously necessary. Whether in the middle Magdalena valley or in other colonisation areas of the rainforest of Latin America, women have little access to land title or to credit (Carmen Diana Deere 1986; Magdelena Leon *et al.* 1987; Halldis Valestrand 1989). In San Lucas, for instance, legal possibilities for access to title and credit would be almost irrelevant to women's opportunities, given the prevailing gender roles and relations. Women colonists have strong views about their needs but limited opportunity to make demands. It is this opportunity which they most require.

21

THE ROLE OF GENDER IN PEASANT MIGRATION

Conceptual issues from the Peruvian Andes[1]

Sarah Radcliffe

INTRODUCTION

Migration in third world areas is usually gender differentiated (Orlansky and Dubrovsky 1978; Young 1979; Khoo et al. 1984; ISIS 1980; Youseff 1979; *International Migration Review* 1984; *Migration Today* 1982). Case studies have demonstrated how male and female migrants play distinct roles in the rural economies they leave, as well as in the labour markets they enter (Radcliffe 1986a; Trager 1984; Pittin 1984; Wilkinson 1983). In this chapter, I argue that gender is a major conceptual tool for understanding the organisation of peasant migration because peasants organise their participation in subsistence-oriented agricultural production and in labour/product markets by using gender-based criteria in the allocation of labour (Young 1980; Radcliffe 1986b). For example, in southern Andean peasant communities, migration experiences are very different for men and women, not only in terms of destinations and the work undertaken, but also in the relationships maintained with the rural households and the work carried out in the domestic unit.

Within Andean peasant households, relationships between members are conditioned by patriarchal norms: female decision-making power and access to resources are set within a context where their participation is under-valued and misrepresented. Given that the domestic units are oriented towards guaranteeing production, the increased commoditisation of the economy leads to the development of multiple class relations by members of the units. In multi-class households where the criteria for deciding who is to migrate, and when, are determined by patriarchal relations between men and women, gender-specific migration patterns result. My interest lies particularly in the selectivity and rationality of migration by peasant women and how this compares with male migration. In order to develop a gender-aware approach to migration, I look at how labour is organised in peasant households in order to depict how the multiplicity of class relations which arise in peasant units is a structured and rational, though patriarchal, process, entailing different outcomes for male and female members. The multi-classness of the peasant household lies behind the insertion of household members in external labour markets and their migration, on either a long- or short-term basis. Under conditions of increasing commodification of the peasant economy, the household is not a simple conglomerate of individuals securing livelihood from the same subsistence and petty commodity base. Rather, individuals undertake different and gender-specific production and labour functions, ranging from subsistence production to migratory wage labour.

One of the aims of this chapter is to examine the contexts and relations for which the household remains a valid concept. I argue that in explaining gender-differentiated migration, the patriarchal household remains a useful concept as it structures the participation of members in external labour markets. The gender inequality fundamental to household constitution means that women's access to the peasant economy sphere is relatively inflexible, and women are more likely to be released into wage labour.

PEASANTRIES AND THE PEASANT HOUSEHOLD

Developing from Chayanovian notions of the domestic unit, analyses of third world peasantries have frequently focused on the peasant household, where maintenance relies upon family labour and a subsistence-oriented rationality. Recent models have confirmed that reliance on family labour results in a changing labour allocation over the demographic cycle of the unit. Criticisms of this model of peasant households point to the economic and political factors in the wider society which structure and condition the units' maintenance (Goodman and Redclift 1981; Schmink 1984). Most significantly, with the expanding monetary economy in many third world areas, the extent of commoditisation of peasant reproduction (through the sale of products and/or labour) has risen, as it becomes increasingly difficult for the peasant unit to guarantee livelihood through subsistence production (e.g. Bernstein 1982). This tends to lead to the proletarianisation of peasant labour and increasing diversification of activities (Lehmann 1982; Gonzales de Olarte 1984; Quijandria and Espinoza 1988), reflecting a 'strategy' of the unit to minimise the risks associated with one activity, or to satisfy subsistence requirements which cannot be met with previous 'traditional' sets of activities.

Thus, we have a picture of a peasant domestic unit, which in order to maintain its subsistence requirements in an increasingly commodified economy, turns to the 'strategy' of diversifying the activities in which its labour participates. Clearly, the particular activities undertaken by the unit depend upon the specific historical and socioeconomic context in which it is located, and on the characteristics of the prevailing labour and product markets. The degree to which the 'simple reproduction squeeze' (Bernstein 1982) commodifies peasant livelihood thus varies from one geographical area to the next, and with time.

ANDEAN PEASANT HOUSEHOLDS AND FEMALE MIGRATION

In the largely agricultural province of Calca in the Peruvian department of Cuzco peasant production continues to be significant and co-exists with smallholdings and larger commercial enterprises. The pattern of peasant migration reveals the complexity of family participation in external labour markets (Radcliffe 1986a). My fieldwork has shown that, broadly speaking, young female labour is sent into urban labour markets to work in domestic service. Education and training is rarely a reason for female migration, although among young men it is more common, with boys being sent as apprentices to urban workshops. Single women tend to move from the village for long periods, and provide a link into the urban centres for rural families (Radcliffe 1987). Once married, women tend to remain in the village and cease to participate in external labour markets. A small minority migrate on a seasonal basis with their husbands, either at the start of the demographic cycle (if they inherit limited resources) or when they are older and offspring have left the unit. Husbands, meanwhile, make regular seasonal moves into external markets throughout their lives, although the ease of obtaining employment once they are elderly is reduced. This pattern of high mobility among the 'fathers' and 'daughters' of the domestic unit (although into distinct labour markets and for different average periods of time), together with lesser mobility among 'sons', and minimal movement among 'mothers' is characteristic of migration among peasant households in the region.

This pattern of migration is based on the links between the rationale of peasant production and the constitution of gendered labour. Due to the patriarchal constitution and reproduction of the 'peasant unit', male and female household members have different opportunities to manage and negotiate the articulation between capitalist and non-capitalist spheres. Hence, they face distinct possibilities in relation to the organisation of labour, and decisions made in the unit have very different effects on them.

The social construction of households, through marriage and inheritance and labour organisation patterns means that short-term decision making regarding the allocation of female labour is carried out by the male head of the household. However, this is an unarticulated practice (*doxa*: Bourdieu 1977), which lies behind all household members' views. Thus, female migration can, at one level, be understood as the outcome of the 'household's' decisions. Yet it is crucial to examine the extent to which this apparently straightforward allocation of labour results from structurally unequal relations between household members and the rationale behind young *campesinas* (peasant women) leaving the villages.

GENDER RELATIONS AND THE HOUSEHOLD

The process of the social construction of gender in the Calca region is one based upon the ideology of complementarity between male and female. Marriage creates not only a new domestic unit, but also a 'complementary' whole, in which both partners gain (equal) adult status. It is worth stressing that inheritance is relatively equal for men and women, both partners bringing property into the new domestic unit upon marriage. Children know each of their parents' plots of land (scattered in different altitude zones), and upon marriage, children of both sexes receive an equal share of their parents' property. The land held by the household is pooled and worked together through the combined labour of husband and wife (and later children), and the resulting product is held by the household as a whole.

Prior to marriage, children and young adults are trained in the appropriate gendered skills, with knowledge of the rights and obligations associated with household, kin networks and community, and their gender roles (Radcliffe 1986a; Bourque and Warren 1981). The gender division of labour for children up to around 6 to 7 years is flexible, although subsequently boys and girls learn different skills. Male children work in the fields to learn agricultural techniques, while girls remain in the household to cook, care for younger siblings, weave and, in some richer households, take part in commercial activity with their mothers. Both male and female children take part in agricultural activity throughout the year: harvest time is particularly busy. As adults too, men and women participate in agricultural production, and a relatively highly defined sexual division of labour exists. Exclusively male tasks include ploughing (with a bullock-drawn plough or footplough), the *aporque* (stage of potato cultivation) and the maize harvest. Female work (such as planting) is taken over by men if necessary, although the reverse does not hold. Other jobs are generally conceived of as being either largely male or female, but do not entail such taboos as does ploughing. In other words, men are socialised to carry out all stages of the agricultural cycle while women are unable to do so without male labour.

The sexual division of labour is based upon the systematic denial of the value and input of female labour, and simultaneously upon the 'centrality' of male labour to the peasant household enterprise. Hence, women's contribution to agricultural activity is both misrepresented and under-estimated by the men, and by social conventions.[2] For example, men, in their representation of the agricultural cycle, deny women's participation in a series of chores which women in fact do carry out (e.g. weeding, harvesting alone) (see Radcliffe 1986a). Also, men do not define certain tasks as agricultural, although women think of themselves as contributing to the agricultural cycle. Cooking for the farm labourers is a case in point, and clearly forms a central part of the work process, yet is excluded by men. Most crucially, female labour is seen by the men to have a lesser value than male, and female participation is perceived as an inherently inferior contribution.

Women and men gain access to, and control over, resources through the creation of domestic units, that is through marriage. In terms of labour, products and prestige, all these can be mobilized once men and women marry. For women, children's labour (especially female) alleviates daily reproductive and productive work. Access to male and female adult labour also rests upon marital status, allowing women to enter labour-exchange circuits. Moreover, married women are responsible for the transformation and storage of agricultural products, and for the common purse. Their knowledge of the amounts of goods in the store rooms, and the socially sanctioned exclusion of men from this sphere, gives women power. However, it is also an onerous and stressful task to organise subsistence over the annual cycle. Decisions to sell products which were originally grown as foodstuffs rest upon both partners, although cash crops are increasingly worked and marketed exclusively by men.

This pattern of labour evaluation and assessment leads to female dependence on male labour. Since agricultural production among these households depends upon, and requires, the mobilisation of extra-household labour. For this reason, Collins (1986: 653) calls the Andean peasant household a subliminal unit, as households rely upon other units for agricultural labour. Labour is mobilised through labour exchanges, called *ayni* or *minka*, in which days of assistance are repaid in kind (labour or products respectively). Both men and women participate in *ayni*, and at one level these networks are independent, in the sense that the conjugal pair create obligations and networks with other adult individuals of their own sex. At another level however, women's *ayni* is structurally subordinate to male *ayni*. When *ayni* is called upon between men and women, women cannot adequately 'repay' male farming labour with their labour and/or cooking. For example, female-headed households (with no adult sons) can call upon male labour of other households for farming their land, yet their reciprocal labour is constituted as inferior and less valuable. The same pattern occurs within the conjugal unit, where women are structurally dependent upon the labour of sons and husbands to work their own land.

GENDER RELATIONS AND MIGRATION

This nexus of the organisation of production and the social evaluation of male and female labour create between them the context for the expulsion of young female labour from the household. As unmarried female members are not useful in extra-household links (only married women take part in *ayni*), they can contribute minimally to the household base. Young female members of the household are 'surplus' to the agricultural base, while their real and potential contribution to labour demands are under-estimated and constrained by the (relatively rigid) sexual division of labour. However, in households where the farm is diversified, female labour may be utilised, as, for example, in richer households with greater numbers of livestock, whose care may be entrusted to unmarried girls. Overall, unmarried female labour is seen as non-essential to the peasant household unit.

In Calca, the availability of single female labour does not enter into male calculations of maintaining peasant production. In other areas of the Andes which have more flexible sexual divisions of labour, such as among the Aymara (Collins 1986, 1988; cf. Laite 1985; Deere 1978), male and female labour can be redistributed among the necessary tasks, and in these cases, female labour is not excluded from peasant production.

Such a gender-unequal peasant production system as found in Calca entails unequal benefit for members of the unit, even if it does consolidate the 'household's' subsistence base over the medium term. Looking particularly at gender inequalities, there are instances where the female position in the unit is weakened by this system. For example, married women bear the burden of extra reproductive and productive work if younger female household members are sent from the village. Young (1980: 77) describes the implications of this situation, in Mexico. 'Although a mother might protest against her daughter being sent away on the grounds that she needed her help in the house and with child minding, her arguments could easily be overruled by those of her husband as manager of the household enterprise.' The same situation occurs in the Peruvian case.

As female offspring's labour does not contribute to the household economy, its expulsion becomes an option for poorer households. This is not necessarily for reasons of high returns on labour as posited in the neo-classical models. When female members enter external labour markets, the return on their labour is not significant, and in comparison to that predicted in neo-classical models (Behrman and Wolfe 1984; Thadani and Todaro 1979; Todaro 1985), it is not the potentially higher wage-earning labour which leaves the peasant unit, but the members whose labour is of minimal value, owing to its gender and peasant origins. Unmarried peasant women work in domestic service and the harvesting of tropical produce, in which the return for their labour is exceedingly low.

The outcome of this process of production organisation and household reproduction is different for male household members. As sons can be utilised in agricultural labour, they represent an opportunity to consolidate the household's economy. By augmenting the quantity of (male) labour in the exchange circuit, their labour is used to realise the value of the household's land as well as to gain more land. Through renting and sharecropping agreements, households with sons can increase the amount of land under cultivation and hence stabilise the subsistence base. Land is redistributed from female-headed households, who, because of lack of male labour, can only gain subsistence through agreements which give them poor returns on their land. Widows share land under these verbal contracts in return for either a rent or half of the harvest. However, neither of these arrangements are equivalent to what they would get if their land were cultivated in a male-headed household.

Over time, therefore, gender differences in migration patterns occur (Radcliffe 1986a). In the case of the study region, nearly a quarter of 'daughters' are absent at any one time, compared with only 14 per cent of sons, and when eldest daughters are considered, this proportion jumps to over one-third. Female migrants move into urban areas for considerably longer periods than males, who dovetail agricultural work with wage labour. Young women remain in urban areas for over 600 days on average, compared with 320 days for male migrants. Women outnumber men in permanent moves (defined as over 2 years' absence) by 4 to 3.

THE CONSOLIDATION OF PATRIARCHAL RELATIONS

Long-term structural disadvantage also occurs for women, as gender-differentiated access to and control over resources arise in the community itself. Male and female access to the peasant base of production becomes differentiated, as land passes progressively from female to male control, for two reasons. Firstly, in inheritance, male children are present in the village to receive parents' land, whereas female offspring are often absent and cannot reinforce rights over land. Female migrants are not automatically disinherited; their absence requires them to choose between returning to the community to oversee sharecropping agreements or implicitly relinquishing their rights. Secondly, as noted above, female-headed households without adult male labour transfer land in temporary contracts to male-headed households. Widows' land is then sometimes subject to appropriation by sharecroppers or renters, as the former are unable to demonstrate active and continuous production, which is the criterion for continued use-rights.

Migration of young female household members is partially explained by their 'superfluity' in the household, as posited in neo-classical theory, but the creation and re-creation of gendered labour and household domestic units explain the type and selectivity of migration among this group.

Adult men tend to be the most 'peasant' in their links with agricultural production and are the most sustained of any household member. In most cases, they can organise their labour time to take part in subsistence production and cash-crop production by calling upon family and extra-household labour to do so. The majority of male adults participate in these activities in the course of a year, while also spending a smaller proportion of days in wage labour. Their structural position in the sexual division of labour and the social construction of households and gender allows them a flexibility and stability which are not experienced by female household members. In a similar way, 'sons' derive their class status from their role in petty commodity and/or subsistence production.

By comparison, adult women are located in the peasant economy to the extent that their labour is recognised as crucial in specific quantities for the daily and generational reproduction of household labour, as well as management of the domestic budget. Their class position derives from the existence of pooled household resources, yet these must be worked with male labour. Once married, women participate in peasant or petty commodity production to the extent that their labour is recognised and required to contribute to household production. For continued peasant and petty commodity production, most households contain one adult female. By contrast, female unmarried labour is less rooted in the peasant economy. Given the sexual division of labour, young female labour does not have a direct and recognised input into subsistence. Consequently, they are likely to enter the external labour markets as wage workers or informal sector workers.

The means by which female labour in the Calca region is reproduced varies greatly with household composition and resource base, as well as the age and family position of the woman. When married, women's labour power is maintained through the combination of subsistence and cash-generating activities in which husbands are engaged, and to which female reproductive and productive work is directed to support. The reproduction of unmarried female labour within the household depends upon the resource base and activity mix of the unit concerned. If the household has a commitment and/or opportunity to develop peasant activities in which female labour is utilised (e.g. herding livestock, weaving, transformation of products), female labour is retained. However, in most households of Calca, the peasant base is not secure enough in the female sphere (which is socially and historically constructed) to reproduce female labour fully, and young women enter migratory flows to external labour markets. Length of absence is conditioned by the nature of inputs made by female labour into the particular mix of activities developed by each individual unit. In turn, this conditions the extent to which female peasant labour is reproduced in the capitalist wage labour sector, and hence contributes to the diversity of class relations centred on the household.

CONCLUSION

Previous models of third world female peasant migration have not adequately addressed the issues of rationale and selectivity which underlie the features of long-term and seasonal moves. An alternative model of peasant migration is offered whereby the peasant household is reconceptualised as a unit of production and reproduction, in which the organisation of production takes place in an increasingly commoditised economy structured by patriarchal-gendered labour relations. This model, by examining the nature of petty commodity production in a capitalist economy, links the selectivity of migration to the patterns of production (agrarian) and reproduction (sexual division of labour, inheritance, resource allocation) of the peasant household.

These factors of labour organisation are centred upon the domestic unit, in which immediate decisions regarding the disposition of labour are made, which follow predictable and regular patterns due to the nature of the articulation of the peasant with the capitalist mode of production. As a result, unmarried peasant women do not have control over their labour in the co-existence of peasant and capitalist production, due to the sociocultural construction of gendered labour in the peasant household.

The criteria for the allocation of female and male labour necessarily change with the market penetration of peasantries[3]. Descriptions of the criteria for labour allocation thus become central to the analysis of gender-differentiated migration and labour disposition. I have suggested that the central elements of such a description include:

1 the sexual division of labour and its degree of flexibility;
2 the social construction of value for different categories of labour;
3 the social constitution of the household and the nature of inter-household links.

All of these elements are inextricably linked to, and arise from, gender differences and the process of the social construction of gender.

Household members enter into distinct class relations due to their varying initial role in the maintenance of peasant production. This process gives rise to a multi-class household in which gender differences in inputs to the peasant sector lead to gender-differentiated commoditised relations. Thus while certain members become petty commodity producers, through their links with agrarian production, other members become channelled into wage labour migration. Changes over time in the gender ratios of migration evolve from transformations in the organisation of household production and in the degree to which household members' labour is reproduced in the peasant economy (subsistence and petty commodity production).

Gender issues are thus crucial to development policies for peasant economies. There is a need for detailed analysis of the dynamics behind the maintenance of poor households, to explain how these factors underlie the gender differences in migration from peasant communities.

APPENDIX
List of workshop participants

Dr Asmah Ahmad, Universitat Kebangsaan Malaysia, Selangor, Malaysia

Miss Julie Ajakpo, University of Swaziland, Kwaluseni, Swaziland

Dr Elizabeth Ardayfio-Schandorf, University of Ghana, Legon, Ghana

Dr Hazel Barrett, Derbyshire College of Higher Education, Derby, UK

Dr Christine Barrow, University of the West Indies, Barbados

Dr Jane Benton, University of Hertfordshire, Watford, UK

Dr Liz Bondi, University of Edinburgh, Scotland

Dr Angela Browne, Coventry, UK

Dr Amriah Buang, Universitat Kebangsaan Malaysia, Selangor, Malaysia

Dr Trudi Bunting, University of Waterloo, Ontario, Canada

Ms Jill Burnett, Oxford University, UK

Dr Gemma Canoves, Autonomous University of Barcelona, Spain

Ms Aditi Chatterji, Oxford University, UK

Dr Nora Chiang, National Taiwan University, Taipei, Taiwan

Mrs Ruvimbo Chimedza, University of Zimbabwe, Harare

Ms Anne Croft, University of Huddersfield, UK

Ms Lindsay Dorney, University of Waterloo, Ontario, Canada

Professor Dennis Dwyer, University of Keele, Keele, UK

Dr K. Maudood Elahi, Jahangirnagar University, Dhaka, Bangladesh

Dr Stephanie Fahey, University of Melbourne, Victoria, Australia

Dr Peggy Fairbairn-Dunlop, University of the South Pacific, Apia, Western Samoa

Ms Debbie Falls, University of Newcastle upon Tyne, UK

Ms Zohra Fatima, Oxford University, UK

Dr M-Dolores Garcia-Ramon, Autonomous University of Barcelona, Spain

Dr Nicky Gregson, University of Sheffield, UK

Mr Hans Guttman, Gothenburg, Sweden

Dr Indra Harry, University of Calgary, Alberta, Canada

Mrs Shahnaz Huq-Hussain, University of Dhaka, Bangladesh

Dr B. Hyma, University of Waterloo, Ontario, Canada

Dr M. Indiradevi, Andhra Pradesh University, Visakhapatnam, A.P., India

Dr Bose Folosade Iyun, University of Ibadan, Nigeria

Dr V. Chandra Jha, Visva-Bharati University, Santinikentan, West Bengal, India

Miss Elizabeth Kasimbazi, University College Swansea, UK

Ms Sara Kinder, University of Durham, UK

Ms Vivian Kinnaird, University of Sunderland, UK

Mrs Kiyomi Kogo, Nakano-Ku, Japan

Ms Sue Langham, University of Newcastle upon Tyne, UK

Ms Alison Lowe, University of Newcastle upon Tyne, UK

Ms Susan McDowell, Newcastle upon Tyne, UK

Dr Fiona Mackenzie, Carleton University, Ottawa, Canada

Mr Edmond Maloba, University of Waterloo, Ontario, Canada

Ms Donny Meertens, University of Amsterdam, The Netherlands

Dr Janet Momsen, University of California-Davis, California, USA

Dr Caroline Moser, London School of Economics, UK

Professor Victoria Mwaka, Makerere University, Kampala, Uganda

Dr N. Nagabhushanam, S.V. University, Tirupati, A.P. India

Dr Cathy Nesmith, Simon Fraser University, British Columbia, Canada

Ms Perez Nyamwange, York University, Ontario, Canada

Ms Lynn O'Malley, Newcastle upon Tyne, UK

Ms Carole Odell, University of Durham, UK

Dr Julie Okpala, University of Nigeria, Nsukka, Nigeria

Ms Grace Okungu, University College Swansea, UK

Professor Gudren Olafsdottir, University of Iceland, Reykjavik, Iceland

Ms Elizabeth Oughton, University of Newcastle upon Tyne, UK

Mr Stephen Price-Thomas, University of Newcastle upon Tyne, UK

Dr Sarah Radcliffe, Royal Holloway and Bedford New College, University of London, Surrey, UK

Mrs Parvati Raghuram, University of Newcastle upon Tyne, UK

Dr Saraswati Raju, Jawaharlal Nehru University, New Delhi, India

Dr Carole Rakodi, University of Wales College of Cardiff, UK

Dr C.H. Ranasinghe, University of Colombo, Sri Lanka

Dr Y. Rasanayagam, University of Colombo, Sri Lanka

Ms Debbie Robinson, Edinburgh, UK

Ms Clare Rohdie, Edinburgh, UK

Dr Colin Sage, Wye College, University of London, Kent, UK

Dr Vidyamali Samarasinghe, The American University, Washington, DC, USA

Ms Penny Smith, University of Newcastle upon Tyne, UK

Ms Montserrat Solsona, Autonomous University of Barcelona, Spain

Dr Janet Townsend, University of Durham, Durham, UK

Dr Rohana Ulluwishewa, University of Sri Jayewardenepura, Nugegoda, Sri Lanka

Ms J. Valentine, James Cook University, Queensland, Australia

Ms Haldis Vallestrand, Universitetet I Tromso, Norway

Mrs Rameswari Varma, University of Mysore, India

Ms Julie Wheelwright, London, UK

Dr A. Wickramasinghe, University of Peradeniya, Sri Lanka

Ms Cara Williams, Clark University, USA

NOTES

2 Household energy supply and women's work in Ghana

This research was undertaken under the International Labour Organisation Energy and Women Project. Their support is hereby acknowledged. I thank in particular, Elizabeth Cecelski for her assistance.

5 The impact of labour-saving devices on the lives of rural African women

Research funding from the Nuffield Foundation is gratefully acknowledged.

1 A study by the Medical Research Council estimates that Gambian women spend, on average, 230 minutes daily in the preparation of *coos* and about 50 minutes each day preparing rice (Roberts *et al.* 1982).

6 The impact of contraceptive use among urban traders in Nigeria

We are grateful to the Fertility Research Unit, University College Hospital, Ibadan, and the Centre for Population and Family Health, Colombia University, New York, for sponsoring part of the research for this chapter.

7 Gender relations in rural Bangladesh

1 A number of *upazilas* form a district (equivalent to a county).
2 Muslim tradition stresses that one cannot please Allah only through worship by abandoning worldly life. By marriage one fulfils obligations mentioned in the Quran and Hadis. The Prophet of Islam urges every Muslim to marry.

8 Seasonality, wage labour and women's contribution to household income in western India

Bob Williams and David Surtees gave me excellent and much-needed advice on the computing and data handling involved in this work, and my thanks also go to Janet Townsend, Janet Momsen, John Lingard and Michael Carrithers for their useful comments.

1 These issues have arisen in the course of a much larger study in which I am

looking at seasonality, household food security and the distribution of subsidised grain.

2 Guyer (1980) shows, using data from Africa, a distinct pattern of male and female seasonality of earning from crop sales rather than employment. In this case male and female earnings are inversely related.

3 Data are not available on intra-household distribution. It is assumed here, therefore that the earnings of the individual are available to the household, although in practice not all income may be used for the common good.

4 The absolute number of households discussed in the study is small. For this reason I have chosen to use a methodology of detailed description and explanation rather than inferential statistical techniques which would be inappropriate.

5 To put these figures in perspective, at this time a school teacher earned approximately Rs600 per month; an office clerk or nightwatchman, Rs400–450.

6 In this chapter I have not distinguished between the different types of work that women do. The most important work is agricultural labouring, but there is a limited amount of indoor domestic work and occasional work on Employment Guarantee Scheme projects.

7 These figures are calculated by first determining the percentage that the women's income accounts for of total income for each household and then calculating the mean across households. This method produces a partially weighted mean that takes into account the level of income of the household.

8 For a discussion of women, their work and views on social status see Bardhan (1985).

9 The data covered the same sample and applied to the 77 households that cultivated some land.

10 It should be noted that these payments cover only casual labour. Annual *balutedar* payments for specialist services such as rope making or other crafts, or grain paid annually to bonded labourers are not shown.

11 In order to determine the cash equivalent to the household I have assumed the price per kilo of grain received is equivalent to the average retail market price of the second-quality grain in the village in the middle of the month it was earned.

9 Invisible female agricultural labour in India

The data upon which this analysis is based was taken from a 1983–4 study in Karoli, Sangli district, Maharashtra, India by Elizabeth Oughton.

10 Assessing rural development programmes in India from a gender perspective

The author gratefully acknowledges the assistance of Sri G.S Ganesh Prasad, Research Investigator, Institute of Development Studies, in the collection and analysis of field data.

1 Mysore district is situated in the southern part of Karnataka State. The total geographical area of the district is 11,947 square kilometres. The district has a population of 2,584,878 according to the 1981 census. Of these 708,275 people live in the urban areas and 1,876,602 live in rural areas. The district consists of 11 taluks (blocks) with 1,641 inhabited villages. The Integrated

Rural Development Programme (IRDP) has been in existence since 1980 and has assisted 54,786 families.

2 Mysore taluk is one of the taluks of Mysore district. It has an area of 810 km^2, 113 inhabited villages and a 1981 population of 162,367 rural dwellers and 476,446 urban inhabitants.

3 Women are supposed to get a fair share in the food-for-work programme, the NREP and the RLEGP. Though there are a number of evaluation studies of these programmes, there is little information about women's participation in the programmes or the impact of the programmes on women's employment and status. The Institute of Development Studies is one of the evaluating agencies in such an exercise. A first look at our findings reveals that sex disaggregated data for women's participation has not been maintained in many cases and where it does exist it shows that women's share is minimal.

4 Marginal farmers and small farmers are classifications based on landholdings. They are among the income group considered to be the target group for development programmes.

5 Right to ownership of land is an important factor influencing the status of women in rural areas (see Bina Agarwal, 1988 'Who sows who reaps' *Journal of Peasant Studies*, 15, 4.)

6 During discussions with bankers while conducting evaluation studies we were told that women beneficiaries maintained the assets better and made more regular loan payments than men beneficiaries.

7 The official booklet on DWRCA says that women beneficiaries of IRDP spent all the increase in income from the investment of the loan on the family, whereas men's increased income is usually partially spent on alcohol, tobacco and entertainment.

11 Access of female plantation workers in Sri Lanka to basic-needs provision

1 Physical Quality of Life Index (PQLI) was devised by officials of the Overseas Development Council. It is a composite index derived statistically from three indicators: infant mortality, life expectancy at age one and literacy rate.

2 There is a distinction between the 'Indian' Tamil plantation workers of Sri Lanka and the Sri Lankan Tamils. The Sri Lankan Tamils and the Sinhalese have been residing in Sri Lanka for more than 2,500 years. The Indian Tamils were brought from south India to the island during the British colonial period in the mid nineteenth century to work in the newly opened plantations, originally devoted to coffee and later to tea production.

3 'Coolie': generally referred to as a native burden carrier or hired labourer. It denotes a subordinate position.

4 An agreement with India provided that of an estimated 975,00 stateless Indian Tamil population, 525,000 were to be given Indian citizenship and repatriated over a 15-year period. Another 300,000 were given Sri Lankan citizenship. Negotiations over the remaining 150,000 people were left for a future date. In 1974, the two governments divided these people on a 50/50 basis.

5 Tea plucker: mainly women plantation workers. They pluck the top two leaves and the bud which is thrown over the shoulder into a basket strapped on the back. Tea is plucked all year round and the same field yields every 7–10 days depending on the flush.

6 'Over kilo' or over poundage refers to the weight of tea plucked over and

above the norm which is set for the women for any given day. Payment per kilo of extra poundage is also stipulated. Cash plucking is when women are paid per kilo with no norms applying. This is additional work beyond the stipulated 8 hours or on holidays. This is practised during peak flush seasons.

7 Estate Medical Practitioner: they are generally referred to as Estate Medical Assistants (EMA) and are in charge of the estate dispensary. They have had training as mid-level health personnel and providers of out-patient treatment for minor ailments and injuries.

14 Gender and the quality of life of households in raft-houses, Temerloh, Pahang, peninsular Malaysia

I acknowledge that this study was inspired by the spirit of Bubolz and others' article on a human ecological approach to quality of life. I also wish to thank the Honours Year Students of the 1987–8 session, Department of Geography, Universiti Kebangasaan Malaysia, for conducting the questionnaire survey as part of their MC4124: Geography and Welfare project.

17 The role of women's organisations and groups in community development

1 Research in 1987 was funded by the British Academy and Hatfield Polytechnic.
2 CIDEM, Directorio de Instituciones Femininas (La Paz: CIDEM, 1986).
3 Popular education, involving the process of conscientisation, is normally understood in Bolivia to include a study of the society's cultural background, the individual's personal welfare (including health and hygiene) and position in society and the way in which this can be improved through organisation and co-operation.
4 This is a figure quoted by CEDOIN in *Bolivia Bulletin*, April 1989, 5: 2.
5 A statement made in the documentary 'Hell to Pay', first shown on Channel 4, 29 May 1988.
6 Quoted from the ECLAC report by Latinamerica Press (Lima: 27 October 1988.
7 Less than 15 per cent of Bolivian women use contraceptives according to *Women in the World: An International Atlas* (Seager and Olson 1986).
8 *The Triple Struggle* (1982) contains a number of lengthy conversations with Latin American women: 'most of the women were involved in some kind of organised co-operative income generating activity or new community development initiatives' (Bronstein).
9 Seager and Olson (1986) give a figure of 49 per cent compared with 24 per cent for Bolivian men.

18 Deconstructing the household

I should like to thank Janet Townsend in particular and also Sarah Whatmore for their help in the preparation of this chapter.

1 Once labour pains commence the expectant woman's mother, or an older woman in the community who has experience as a *partera* (midwife), is summoned to facilitate delivery which always takes place at home. Few households have beds, so the woman usually gives birth on a dirt floor covered

by a blanket. Often the woman's partner or father participates, holding her in an upright position. It is believed that the uterus can move up through the woman's body and emerge at the mouth, so a belt (chumpi) is tightened around the body before the placenta follows the baby. This often causes the placenta to be retained inside the body with the baby still attached via the umbilical cord. This can be left uncut for days or is simply cut with an unsterilised knife or broken with a stone. This can produce septicaemia in the new-born child as well as in the mother when the placenta is retained by the body. The doctor attached to a local hospital informed me that from the problems of infection resulting from retention of the placenta and tearing of cervical tissue, he had attended at the death of over 20 women in just 6 months.

19 Women's roles in colonisation

A Spanish version of this text was presented at the 46th International Congress of Americanists, 4–8 July 1988 in Amsterdam, Holland, and published in *Colombia Amazonica*, 3: 2, December 1988.

For the fieldwork and interpretation of the data collected we had the invaluable assistance of Zoraida Ordonez, a social worker specialising in rural development. To the peasant women of the San Jose del Guaviare and El Retorno settlements goes our gratitude for the dynamic way in which they co-operated with this research and for the trust they bestowed upon us. But more than gratitude it is necessary to reaffirm here a commitment to bring back results that can be discussed by them in light of their own analysis of reality, in order to find through dialogue and their recommendations, adequate ways to continue promoting their full participation in regional development.

1 The 16 small villages include a population of approximately 500 families, of which the sample represents 15 per cent. Estimates of the total pioneer population of Guaviare (excluding the floating population of hired hands) fluctuate between 3,500 and 6,000 families in an area of approximately 800,000 hectares.
2 The settlement became 'official' with 181,000 hectares taken from the Forest Reserve through Inderena Resolution of 1969 and the subsequent arrival of INCORA and Caja Agraria, which, however, in the first few years only reached a minute proportion of the region's settlers.
3 Gonzalez 1980; Caceres 1980; Deere and Leon 1983; Hamilton 1986; Lisansky 1979 and Hecht n.d.
4 It is interesting to compare this with other survival crisis situations in which women with small children participate more actively in acts geared to radically transforming their circumstances, for example, rural or urban land invasions. Under 'normal' conditions, however, it appears that the rate of women's involvement increases as they, and their children, get older (CIPAF 1985).
5 The results of the research undertaken in Guaviare differ in this sense from those found by Janet Townsend in some settlement projects in the Magdalena Medio and the Atlantic coast. According to her, female participation diminishes strongly after the first settlement stage; once the farm is formed she is practically ousted from productive chores. Townsend found no involvement of women in home gardens or with livestock. To what extent is this difference due to the different source areas of the migrants, in Townsend's case the Atlantic Coast of Colombia? (Townsend, Chapter 20).

21 The role of gender in peasant migration

I wish to thank Cathy Nesmith and Stephanie Fahey for discussions around the issues in this chapter, and comments on earlier drafts.

1 A longer version of this chapter appears in *Review of Radical Political Economics*, Special Issue on 'Women and Development', 1991.

2 Women's participation is also systematically denied in the literature. Boserup (1970: 187–8) characterised the Latin American region as a male-farming system (although detailed observation in the Andes has subsequently disproved this claim), and census data under-enumerates female work in agriculture (Deere and Leon 1982).

3 It could also be hypothesised that labour allocation procedures would be transformed with a change in supra-household resource mobilisation, the other half of the equation (see Collins 1988).

REFERENCES

PART I INTRODUCTION

1 Geography, gender and development

Brydon, L. and Chant, S. (1989) *Women in the Third World: Gender Issues in Rural and Urban Areas*, London: Edward Elgar.

Christopherson, S. (1989) 'On being outside the project', *Antipode*, 21, 2: 83–9.

Foord, J. and Gregson, N. (1986) 'Patriarchy: towards a reconceptualisation', *Antipode*, 18, 2: 186–211.

Glick, T. S. (1986) 'History and philosophy of geography', *Progress in Human Geography*, 10, 2: 267–77.

Mohanty, C. T. (1991) 'Under Western eyes: feminist scholarship and colonial discourses', in C. Mohanty *et al.* (eds) *Third World Women and the Politics of Feminism*, Bloomingdale: Indiana University Press.

Momsen, J. (1991) *Women and Development in the Third World*, London: Routledge.

Momsen, J. and Townsend, J. (1987) *Geography of Gender in the Third World*, London: Hutchinson.

Probyn, E. (1990) 'Travels in the postmodern: making sense of the local', in L. Nicholson (ed.) *Feminism/Postmodernism*, London: Routledge.

Spivak, G. (1988) 'Can the subaltern speak?', in C. Nelson and L. Grossburg (eds) *Marxism and the Interpretation of Cultures*, Chicago: University of Illinois Press.

Townsend, J. (1991) 'Towards a regional geography of gender', *Geographical Journal*, 157, 1: 25–35.

United Nations (1991) *The World's Women 1970–1990: Trends and Statistics*, New York: United Nations.

PART II AFRICA

Introduction

Ardayfio-Schandorf, E. (1986) *The Rural Energy Crisis in Ghana: Implications for Women's Work and Household Survival Strategies*, Geneva: ILO.

Dankelman, E. and Davidson, J. (1987) *Women and Environment in the Third World: Alliance for the Future*, London: Earthscan Publications.

World Commission on Environment and Development (1987) *Our Common Future*, Oxford: Oxford University Press.

2 Household energy supply and women's work in Ghana

Agarwal, A. (1985) *Energy and the Third World Subsistence Sector*, Nairobi: ELCI.
Ardayfio, Elizabeth (1986) *The Rural Energy Crisis in Ghana: Its Implications for Women's Work and Household Survival*, Geneva: International Labour Organisation.
Ardayfio-Schandorf, Elizabeth (1983) 'Household energy utilisation patterns in south-western Nigeria', *Bulletin of the Ghana Geographical Association*, new series.
—— (1987) 'Rural energy supply: alternative strategies', paper given at the second National Symposium on Renewable Energy, University of Ghana.
Barnard, G. W. (1987) *Woodfuel in Developing Countries*, Chichester: John Wiley and Sons.
Barnor, L. (1985) 'Energy situation in Ghana', paper given at the workshop on policy implications of rural energy in developing countries, The Hague: ILO.
Boserup, E. (1970) *Women's Role in Economic Development*, London: Allen & Unwin.
Bukh, J. (1979) *The Village Woman in Ghana*, Upsala: Scandinavian Institute of African Studies.
FAO (1986) *African Agriculture: the Next 25 Years*, Rome: Atlas of African Agriculture.
Ghai, D. (1986) *Energy and Rural Women's work*, vol. 1, Geneva: ILO.
Government of Ghana (1970) Ghana Population Census, Accra.
—— (1988) Ministry of Fuel and Power, Accra.
Okali, C. and Kotey, R. A. (1975) *Okokooso: A Resurvey*, Technical Publication Series, no. 15, ISSER, Legon: University of Ghana.
Pala, A. and Seidman, A. (1976) *A Proposed Model of the Status of Women in Africa*, Legon: University of Ghana.
Parikh, J.K. (1980) *Energy Systems and Development*, Delhi: Oxford University Press.
Pogucki (1955) 'Land tenure in Ghana', mimeo.
Stevens, Y. (1986) *Energy Efficient Technologies for Increasing Incomes in Rural Food Processing in West Africa*, Geneva: ILO.
Udo, R. K. (1982) *The Human Geography of Tropical Africa*, Oxford: Clarendon Press.

3 Women's role and participation in farm and community tree-growing activities in Kiambu district, Kenya

Bernard, J. (1981) *The Female World*, New York: Free Press.
Bronkensha, D. and Alfonso, P. C. (1984) *Fuelwood, Agro-forestry and Natural Resources Management: the Development Significance of Land Tenure and Other Resource Management Utilization Systems*, New York: Institute for Development Anthropology.
Central Bureau of Statistics (1979) Census, Nairobi: Government of Kenya.
—— (1983) Land Categorisation, Nairobi: Government of Kenya.
Dankelman, I. and Davidson, J. (1988) *Women and Environment in the Third World: Alliance for the Future*, London: Earthscan Publications.
FAO (1983) *Rural Women, Forest Outputs and Forestry Projects*, Rome: FAO.
—— (1985a) *Tree Growing by Rural People*, Rome: FAO, Forestry Paper 64.
—— (1985b) *Intensive Multiple-use Forest Management in the Tropics: Analysis of Case Studies in India, Africa, Latin America and the Caribbean*, Rome: FAO.
—— (1986) *Forestry Extension Organisation*, Rome: FAO.

Fortmann, L. (1986) 'Women and subsistence forestry: cultural myths form a stumbling block', *Journal of Forestry*, 84, 7: 39–42.
Hoskins, M. W. (1979) *Women in Forestry for Local Community Development: a Programme Guide*, Washington: USAID.
March, K. S. and Taqqu, R. L. (1986) *Women's Informal Associations in Developing Countries*, Boulder: Westview.
Mathaai, W. (1988) *The Green Belt Movement*, Nairobi: National Council of Women in Kenya.
Pala, A. O. (1976) *African Women in Rural Development: Research Trends and Priorities*, Nairobi: Institute of Development Studies, paper 12.
Sanwal, M. (1988) 'Community forestry: policy issues, institutional arrangements and bureaucratic reorganisation', *Ambio*, 17, 5: 342–6.
Shiva, V. (1987) 'Forestry myths and the World Bank', *The Ecologist*, 17, 4/5: 142–9.
Staudt, K. (1985) 'Women's political consciousness in Africa: a framework for analysis', in J. Monson and M. Kalb (eds) *Women as Food Producers in Developing Countries*, Los Angeles: UCLA, African Studies Centre.
Tostensen, A. and Scott, J. G. (1987) *Kenya: Country Study and Norwegian Aid Review*, Bergen: DERAP Publication 224.
Uphoff, N. T. (1986) *Local Institutional Development: an Analytical Source Book with Cases*, West Hartford: Kumarian Press.
Vollers, M. (1984) 'Healing the ravaged land', *International Wildlife*, 18, 1: 5–11.
Wanyande, P. (1987) 'Women's groups in participatory development: Kenya's development experience through the use of harambee', *Development: Seeds of Change*, 2, 3: 94–102.
Williams, P. J. (1982) 'Women and forest resources: a theoretical perspective', in M. Stock, J. E. Force and D. Ehrenreich (eds), *Women and Natural Resources, International Perspectives*, Moscow, Idaho: Forest, Wildlife and Range Experiment Station, University of Idaho.

4 Agricultural production and women's time budgets in Uganda

Bradley, P. N. (1991) *Woodfuel, Women and Woodlots, vol. 1: the Foundations of a Woodfuel Development Strategy for East Africa*, London: Macmillan.
Momsen, Janet Henshall (1991) *Women and Development*, London: Routledge.
Mwaka, Victoria M. (1988) 'Irrigation farming in Iganga and Kamuli districts in Uganda', unpublished preliminary research results, Kampala: Department of Geography, Makerere University.
Okeyo, A. P. (1980) 'The Jolua equation: land reform equals lower status for women', *Ceres*, 75.
Republic of Uganda, Ministry of Planning and Economic Development (1987) *Rehabilitation and Development Plan 1987/1988 to 1990/1991*, Kampala.
United Nations High Commission for Africa (1984) *Integration of Women into African Development*, New York.

5 The impact of labour-saving devices on the lives of rural African women

Adeyokunnu, T. (1984) 'Women and rural development in Africa', in UNESCO, *Women on the Move*, Paris: UNESCO.
Barrett, H. and Browne, A. (1989) 'Time for development', *Scottish Geographical Magazine*, 105: 4–11.

Carr, M. (1985) 'Technologies for rural women: impact and dissemination', in I. Ahmed (ed.) *Technology for Rural Women*, London: Allen & Unwin.
Catholic Relief Services (CRS) (1988a) *Annual Public Summary of Activities in The Gambia, 1987*, The Gambia: CRS.
—— (1988b) Personal communication.
Intermediate Technology (IT) (1985) *Tools for Agriculture*, London: Intermediate Technology Publications.
Ivan-Smith, E. *et al.* (1988) *Women in Sub-Saharan Africa*, Minority Rights Group Report 77.
Mabogunje, A. (1980) *The Development Process*, London: Hutchinson.
Ministry of Planning and Economic Development (1987) Agricultural Statistics, Kampala: Government of Uganda.
Momsen, J. and Townsend, J. (eds) (1987) *Geography of Gender in the Third World*, London: Hutchinson.
Rice, J. E. (1979) 'Agricultural extension services', unpublished monograph for UN Rural Development Project, The Gambia.
Roberts, S. B. *et al.* (1982) 'Seasonal changes in activity, birth weight and lactational performance in rural Gambian women', *Transactions of the Royal Society of Tropical Medicine and Hygiene*, 76: 668–78.
Rogers, B. (1980) *The Domestication of Women*, London: Kogan Press.
Sandu, R. and Sandler, J. (eds) (1986) *The Technology and Tools Handbook*, London: Intermediate Technology Publications.
Todaro, M. P. (1977) *Economic Development in the Third World*, London: Longman.
Trenchard, E. (1987) 'Rural women's work in sub-Saharan Africa and the implications for nutrition', in J. Momsen and J. Townsend (eds) *Geography of Gender in the Third World*, London: Hutchinson.
United Nations (1985) *Forward-looking Strategies for the Advancement of Women*, New York: UN.
UNCTAD (1988) *The Least Developed Countries, 1987 Report*, New York: UN.
UNIFEM (1988) 'Sorghum and millet decortication and flour milling project, final report', unpublished document, Gambia Women's Bureau.
World Bank (1975) *Rural Development*, Sector Policy Paper, Washington, DC: World Bank.

6 The impact of contraceptive use among urban traders in Nigeria

Dow, Thomas E. Jr. (1977) 'Breastfeeding and abstinence among the Yoruba', *Studies in Family Planning*, 18.
Farouq, G. M. and Adeokun, L. A. (1976) 'Impact of a rural family planning programme in Ishan, Nigeria, 1969–1972', *Studies in Family Planning*, 7, 6: 165–8.
IPPF, (1985) 'Two hundred and fifty million couples using contraception', *Review of Population and Development*, 12, 2.
Iyun, Folosade B. (1985) 'Ibadan market documentation study, phase 1', prepared for the Fertility Research Unit, College of Medicine, University of Ibadan in collaboration with Centre for Population and Family Health, Colombia University, New York.
Iyun, Folosade B., Oke, E. A. and Matanmi, O. O. (1987) 'Analysis of baseline survey for the Ibadan low-cost market-based health-care delivery', Ibadan: Fertility Research Unit, University of Ibadan, Nigeria.
Knodet, John and Pitaktepsombati, Pichit (1973) 'Thailand fertility and family planning among rural and urban women', *Studies in Family Planning*, 4, 9: 243–53.

Nigerian Fertility Survey (1981–2), Vol. 1 (1984) 'Methodology and findings', *World Fertility Survey*, 105–19.
Ojo, P. A. (1984) 'Family planning at the UCH, Ibadan 1965–1982' *Planners Forum Magazine*, Pathfinder Project 1, 1: 1–4.
Oni, Gbolahan A. (1986) 'Contraceptive knowledge and attitudes in urban Ilorin, Nigeria', *Journal of Biosoc. Sci.*, 18: 273–83.
Osuntokun, B. O. (1987) 'Health care delivery and management in a teaching hospital in tropical Africa', in O. A. Erinosho and M. O. Akindele (eds) *Proceedings of the Roundtable on Health held at the Aro Neuropsychiatric Hospital, Abeokuta, Nigeria*, prepared for the African Development Foundation, Washington, DC.
Oyediran, M. A. and Ewumi, E. O. (1976) 'Profile of family planning clients at the family health clinic, Lagos, Nigeria', *Studies in Family Planning*, 7, 6.
Rizk, Hanna (1977) 'Trends in fertility and family planning in Jordan', *Studies in Family Planning*, 8, 4.
Saba, S. L. (1987) 'Health care and health services in Nigeria', in O. A. Erinosho and M. O. Akinlele (eds) *Proceedings of the Roundtable on Health held at the Aro Neuropsychiatric Hospital, Abeokuta, Nigeria*, prepared for the African Development Foundation, Washington, DC.
Vagale, L. R. (ed.) (1972) 'Anatomy of traditional markets in Nigeria: focus on Ibadan City', Ibadan: The Polytechnic.

PART III SOUTH ASIA

Introduction

Ahsan, R. M. and Hussain, S. H. (1987) 'A note on female and child labour among the urban poor in Bangladesh', *CUS Bulletin*, no. 13, Bangladesh: CUS Department of Geography, University of Dhaka.
Chadna, R. C. (1967) 'Female working force of rural Punjab: 1961', *Manpower Journal*, 2: 48–51.
Chopra, L. (1979) 'Female participation in the three crop regions of India: an inter-temporal study of rural India between 1951, 1961 and 1971', occasional paper, India: Centre of Study of Regional Development, Jawaharlal Nehru University.
Elahi, K. M. (1982) 'Anthropological indepth study of women's roles and demographic issues in Bangladesh: the indepth questionnaire', Dhaka: ILO/UNFPA.
—— (1987) 'Rural fertility and female economic activity', Dhaka: ILO/UNFPA.
Gosal, G. S. (1961) 'The regionalism of sex composition of India's population', *Rural Sociology*, 26: 122–37.
Kundu, A. (1986) 'Inequity in education development: issues in measurement, changing structure and its socio-economic correlates with special reference to India', in Moonis Raza (ed.) *Educational Planning: a Long-Term Perspective*, New Delhi: Institute of Educational Planning and Administration.
Mitra, A. (1978) 'Employment of Women', *Manpower Journal*, 14, 1–16.
—— (1979a) *Implications of Declining Sex Ratio in India's Population*, New Delhi: Allied Publishers.
—— (1979b) *Status of Women: Employment and Literacy*, New Delhi: Allied Publishers.
Mittal, M. (1986) 'Women in the informal urban economy, a geographical study

of Meerut City', unpublished Ph.D. thesis, Haryana: Department of Geography, University of Kurukshetra.

Mukerji, A. B. (1971) 'Female participation in agricultural labour in Uttar Pradesh: spatial variation', *National Geographer*, 6, 13–18.

Nagia, S. (1983) 'Work status of female migrants before and after migration, a case study of Salem City in Tamil Nadu', *Annals of the National Association of Geographers*, 2, 2: 39–53.

Nagia, S. and Samuel, M. J. (1983) 'Determinants and characteristics of female migration: a case study of Salem City in Tamil Nadu, India', *Population Geography*, 5, 1–2: 34–43.

Nayak, D. K. and Ahmad, A. (1984) 'Female participation in economic activity: a geographical perspective with reference to rural areas in India', *The Indian Geographical Journal*, 59, 262–7.

Nuna, S. (1986) *Spread of Female Literacy in India, 1901–1981*, New Delhi: National Institution of Educational Planning and Administration.

Patnaik, I. (1989) 'Women in the urban informal sector', *Other Sociology*, 1, 4: 1–3.

Raju, S. (1981) 'Sita in the city: a sociogeographical analysis of female employment in urban India', *Discussion Paper 68*, Syracuse: Department of Geography, Syracuse University.

—— (1982) 'Regional patterns of female participation in the labor force in India', *Professional Geographer*, 34, 42–9.

—— (1984) 'Female participation in the urban labor force: a case study of Madhya Pradesh, India', working paper, Women and Development, Ann Arbor: Michigan State University.

—— (1987) 'A socio-geographical analysis of female participation in the labour force in urban India: Madhya Pradesh as an example', *Asian Profile*, 15, 247–67.

—— (1988) 'Female literacy in India: the urban dimension', *Economic and Political Weekly*, 44, 57–64.

—— (1989) 'Research on gender, some conceptual and theoretical issues', paper presented at the Workshop on Gender and Development, Newcastle upon Tyne, 16–21 April.

—— (1991) 'Female labor, poverty and regional culture: the Indian experience', paper presented at the Seventeenth Pacific Science Congress, Honolulu, Hawaii, 27 May-2 June, 1991.

Ranasinghe, P. C. H. (1989) 'Female social mobility and education in Sri Lanka: a critical evaluation', paper presented at the Workshop on Gender and Development, Newcastle upon Tyne, 16–21 April.

Satish, M. (1984) 'Tertiarisation and female participation in the Indian urban economy 1971', paper presented in a research seminar on Third World Urbanisation and the Household Economy, University Sans Malaysia, Penang, 15–16 August.

Swarnkar, G. P. (1985) 'Rural women working population of Mayhya Pradesh – a case study in population geography', unpublished Ph.D. thesis, Madhyra Pradesh.

Wickramasinghe, A. (1987a) 'The socio-economic survey of the rural women: the innovation for income generating activities for the integrated rural development project (IRDP): Kegalle district', Colombo: Ministry of Plan Implementation.

—— (1987b) 'Income generation for rural women in the Kegalle district under the integrated rural development project', paper presented at the Department of Geography, staff seminar series, University of Paradeniya, 17 June.

—— (1990a) 'Assessing the situation of women in system B: a prerequisite to planning a training programme', Sri Lanka: Mahaweli Regional Training Centre.

—— (1990b) 'Evaluation of women's income generating activities under the integrated rural development project in the Kegalle district', Sri Lanka: IRDP.

7 Gender relations in rural Bangladesh

BBS (1976) *Bangladesh Population Census, Bulletin 1, Union Statistics 1974*, Dhaka: Bangladesh Bureau of Statistics, Government of Bangladesh.

—— (1984) *Bangladesh Population Census 1981, Analytical Findings and National Tables*, Dhaka: Bangladesh Bureau of Statistics, Government of Bangladesh.

—— (1985) *Bangladesh Population Census 1981, Community Tables, Thana Series: Tangail and Comilla*, (2 vols) Dhaka: Bangladesh Bureau of Statistics, Government of Bangladesh.

Elahi, K. M. (1982) *Anthropological Indepth Study of Women's Roles and Demographic Issues in Bangladesh: the Indepth Questionnaire*, Dhaka: ILO/UNFPA.

—— (1987) *Rural Fertility and Female Economic Activity in Bangladesh*, Dhaka: ILO/UNFPA

Government of Pakistan (1964) *Census of Pakistan 1961, District Census Reports: Comilla and Mymensingh* (2 vols) Karachi: GOP.

Khan, M. E., Dastidar, S. K. G. and Bairathi, S. (1985) *Health Practices in Uttar Pradesh: a study of discrimination against women*, Baroda: Operations Research Group.

Rabbani, A. K. M. and Hussain, S. (1984) 'Level of fertility and mortality from the national sample vital registration system of the BBS, in Population Development Planning Unit', *Recent Trends in Fertility and Mortality in Bangladesh: Proceedings of National Seminar*, Dhaka: Planning Commission, Government of Bangladesh.

8 Seasonality, wage labour and women's contribution to household income in western India

Bardhan, K. (1985) 'Women's work, welfare and status, forces of tradition and change in India', *Economic and Political Weekly*, 20: 50.

Guyer, J. (1980) *Household Budgets and Women's Incomes*, African Studies Centre working paper 28, Boston University.

Jiggins, J. (1986) 'Women and seasonality: coping with crisis and calamity', *Institute of Development Studies*, Bulletin 17,3.

Lipton, M. (1983) *Labour and Poverty*, World Bank Staff working paper 16, Washington, DC.

Raikes, R. (1981) 'Seasonality in the rural economy of Tropical Africa', in R. Chambers (ed.) *Seasonal Dimensions of Rural Poverty*, London: Pinter.

Rosenzweig, M. (1987) 'Rural wages, labour supply and land reform: a theoretical and empirical analysis', *American Economic Review*, 68, 3.

Ryan, J. and Ghodake, R. D. (1980) 'Labour market behaviour in rural villages of south India: effects of season, sex and socio-economic status', Hyderabad: ICRISAT Economics Progress Report 15.

9 Invisible female agricultural labour in India

Agarwal, B. (1985) 'Work participation of rural women in the third world: some data and conceptual biases', *Economic and Political Weekly*, 20: 51–2.

Anker, R. (1983) 'Female labour force participation in developing countries: a critique of current definitions and data collection methods', *International Labour Review*, 122: 6.
Boserup, E. (1970) *Women's Role in Economic Development*, New York: St Martins Press.
Deere, C. D. and Leon de Leal, M. (1982) *Women in Andean Agriculture*, Geneva: ILO.
Government of India (1971) 'Indian Census through a hundred years', Part I, Census Centenary Monograph, no. 2.
—— (1981) Census of India 1981, New Delhi: Government of India.
Hart, G. (1980) 'Patterns of household labour allocation in a Javanese village', in Binswanger *et al. Rural Household Studies in Asia.*
Kahlon, A. S., Miglani, S. S. and Mehta, S. K. (1973) *Studies in the Economics of Farm Management in Ferosepur District (Punjab): Report for the Year 1967–1968*, New Delhi: Directorate of Economics and Statistics, Ministry of Agriculture.
Sen, I. (1988) 'Class and gender in work time allocation', *Economic and Political Weekly*, 23, 33: 1702–6.

10 Assessing rural development programmes in India from a gender perspective

Committee on Status of Women in India (1975) *Towards Equality*, New Delhi: Indian Council of Social Science Research.
Dantwala, M. L. (1985) 'Garibi Hato: Strategy Options', *Economic and Political Weekly*, 16 March.
Dantwala, M. L., Gupta, R. and D'Souza, K. C. (eds) (1986) *Asian Seminar on Rural Development: the Indian Experience*, Oxford and IBH.
Government of India Planning Commission (1980) *VI Five Year Plan, 1980–1985*, New Delhi: Government of India.
—— (1985) *VII Five Year Plan, 1985–1990*, vol. I and vol. II, New Delhi: Government of India.
Government of India, Department of Rural Development (1980) *Manual on Integrated Rural Development Programme*, New Delhi: Government of India.
Hirway, Indira (1985) 'Gariki Hatao: Can IRDP do it?', *Economic and Political Weekly*, 30 March.
Indian Journal of Agricultural Economics (1985) 40: 3.
Jain, Devaki and Bannerjee, Nirmala (1986) *Women in Poverty: the Tyranny of the Household*, New Delhi: Vikas.
Krishnaraj, Maithreyi (1988) *Women and Development: Analytical Perspectives*, Bombay: Research Centre for Women's Studies, SNDT University.
Kumad, Sharma (1985) *Women and Development: Gender Concerns*, occasional paper, New Delhi: Centre for Women's Development Studies.
Mehra, Rekha and Saradamoni, K. (1983) *Women and Rural Transformation*, New Delhi: Indian Council of Social Science Research.
Neeradesai and Krishnaraj, Maithreyi (1987) *Women and Society in India*, New Delhi: Ajanta Publications.
Parthasarathy, G. (1988) 'Public policy and trends in women and development: rural India', paper presented at the Workshop on Women in Agriculture, Centre for Development Studies, Trivandrum, India.
Rao, V. M. and Erappa, S. (1987) 'IRDP and rural diversification: a study in Karnataka', *Economic and Political Weekly*, 22: 52.

Rath, N. (1985) 'Garibi Hatao: Can IRDP do it?', *Economic and Political Weekly*, 9 February.
Sen, Gita and Grown, Caren (1985) *Development, Crises and Alternative Visions: Third World Women's Perspectives*, New Delhi: DAWN.
Venkatareddy, K. (1988) *Rural Development in India: Poverty and Development*, Bombay: Himalaya Publishing House.

11 Access of female plantation workers in Sri Lanka to basic-needs provision

Bastian, Sunil (1983) *Two Leaves and a Bud: Tea Industry Since Nationalising*, Colombo: Centre for Society and Religion.
Buvinic, M. *et al.* (1988) 'Weathering economic crisis: the crucial role of women in health', in D. Bell and M. Reich (eds) *Health, Nutrition and Economic Crisis: Approaches to Policy in the Third World*, Boston: Auburn House Publishers.
Caldwell, J., Gaminiratne, K. H. W., Caldwell, Pat, de Silva, Soma, Caldwell, Bruce, Weeraratne, Nanda and Silva, Padmini (1987) 'The role of traditional fertility regulations in Sri Lanka', *Studies in Family Planning*, 18: 1.
Central Bank (1984) *Report on Consumer Finance and Socio-economic Survey, 1981/82*, Colombo.
Chatopadya, Haraprasad (1979) *Indians in Sri Lanka: a Historical Study*, Calcutta: O.P.S. Publishers.
Department of Census and Statistics (1978) *World Fertility Survey*, First Report 1975, Colombo.
—— (1981) *Census of Population*, Sri Lanka, Colombo.
—— (1987) *Vital Statistics 1967–1980 Sri Lanka*, Colombo.
—— (1988) *Statistical Pocket Book*, Colombo.
Edirisinghe, N. (1986) *The Food Stamp Program in Sri Lanka: Costs, Benefits and Policy Options*, Washington DC: International Food Policy Research Institute.
Fernando, S. (1985) 'Improvement in education among women in Sri Lanka over the 1977–1983 years', *Economic Review*, People's Bank Colombo.
Gnanamuttu, G. A. (1977) *Education and the Indian Plantation Workers in Sri Lanka*, Colombo.
Guyer, J. (1980) *Household Budgets and Women's Incomes*, working paper 28: Boston: African Studies Center, Boston University, Massachusetts.
Janatha Estates Development Board (1984) *Report of the Orientation Seminar for Estate Superintendents on formulation of Nutrition Intervention Programmes*, Colombo.
—— (1986) *Housing Needs Assessment Synopsis*, Social Development Division, Colombo.
Jayawardena, Kumari and Jayaweera, Swarna (1985) 'The integration of women in development planning: Sri Lanka' in N. Heyzer (ed.) *Missing Women: Development Planning in Asia and the Pacific*, Kuala Lumpur: Asian Pacific Development Centre.
Jayaweera, Swarna (1985) *Women in Employment in UN Decade for Women: Progress and Achievements of Women in Sri Lanka*, Colombo: ENWOR.
Kampschoer, V., Leyen, T., van den Berk, E. and van den Brekel, M. de Jong (1983) *The Road to Health is Paved with Tea Leaves: Medical Care on Sri Lankan Estates*, Amsterdam: Sri Lanka Information Group.
Kumar, S. (1977) *Role of Household Economy in Determining Child Nutrition at Low Income Levels: a Case Study of Kerala*, Ithaca, NY (mimeo).
Kurien, R. (1981) *Women Workers in Sri Lanka Plantation Sector: an Historical and Contemporary Analysis*, Women, Work and Development no. 5, Geneva: ILO.

Langford, C. M. (1982) *Fertility and Tamil Estate Workers in Sri Lanka, World Fertility Survey Scientific Report*, 31, Voorburg: International Statistical Institute.
Lipton, M. (1983) *Poverty, Undernutrition and Hunger*, Washington DC: World Bank Staff, working paper 597.
Marga Institute (1983) *Evaluation of the UNICEF Assisted Programme in the Estate Sector 1979–1983*, Colombo.
Ministry of Plan Implementation (1982) *Evaluation Report on the Food Stamp Scheme*, Colombo: Food Nutrition Policy Planning Division.
Morris, D. (1979) *Measuring the Conditions of the World's Poor: the Physical Quality of Life Index*, New York: Pergamon.
Peiris, G. H. (1988) 'Changing prospects of the plantation workers of Sri Lanka', paper presented at the International Workshop on Economic Dimensions of Ethnic Conflict, Centre for Ethnic Studies, Kandy, Sri Lanka, 13–15 July.
Ratnayake, Kanthi *et al.* (1984) *Fertility Estimates for Sri Lanka Derived from the 1981 Census*, Colombo: Aitken Spence.
Rote, R. (1986) *A Taste of Bitterness: the Political Economy of Tea Plantations in Sri Lanka*, Amsterdam: Free University Press.
Samarasinghe, V. (1985) 'An evaluation of the differential impact of public policy on spatial inequalities in Sri Lanka since 1977', paper presented at International Workshop on Spatial Inequalities, Kano, Nigeria, 16–20 September.
—— (1986) *Women, Wage Employment and Education in Sri Lanka*, United Nations: University of Tokyo.
Sen, A. K. (1980) *Levels of Poverty: Policy and Change*, Washington, DC: World Bank Staff, working paper 401.
Wesumperuma, D. (1986) *Indian Immigrant Plantation Workers in Sri Lanka: a Historical Perspective 1880–1901*, Colombo: Vidyalankara Press.
World Bank (1988) *World Development Report*, 1988, Oxford: Oxford University Press.

12 Women as agents and beneficiaries of rural housing programmes in Sri Lanka

Chandra, H. Soysa (1976) 'The role of housing in improving rural habitat' (mimeo).
Chandrasena, U. A. (1986) *Impact of Rural Housing Programmes on Rural Development in Sri Lanka*, Research Publication 6, Colombo: Centre for Housing, Planning and Building.
Government of India (1981) *Reports of the Census and Statistics, 1981*, New Delhi: Government of India.
IRED (International Innovations and Networks Agency for International Development) (1987) *Aspects of Support Based Rural Housing in the Galle, Matara, Jaffna and Puttalam Districts of Sri Lanka*, Colombo: IRED.
Marga Institute (1986) 'Housing development in Sri Lanka' (mimeo).
Ministry of Local Government Housing and Construction (MLGHC) (1984) *Implementation Guidelines: Rural Sub-Programme*, Colombo: Government of Sri Lanka.
Moser, Caroline and Peake, Linda (eds) (1987) *Women, Human Settlements and Housing*, London: Tavistock.
Rasanayagam, Y. (1988) 'Housing as a point of entry to rural development in Sri Lanka: an appraisal', *University of Colombo Review*, 8.
Robson, D. G. (1984) *Aided Self-help Housing in Sri Lanka*, London: ODA.

Sirivardana, Susil (1982) 'Housing in rural development', (mimeo).
Weerasinghe, G. M. S. (1987) 'Women's participation in rural development', (mimeo).

13 Women's roles in rural Sri Lanka

Boserup, E. (1970) *Women's Role in Economic Development*, New York: St Martin's Press.
ILO (1976) *Report on Employment, Growth and Basic Needs*, Geneva: ILO.
Leach, E. R. (1960) 'The Sinhalese of the dry zone of Sri Lanka', in G. P. Murdock (ed.) *Social Structure in Southeast Asia*, Chicago: Quadrangle.
Palmer, I. (1978) 'The integration of Women in agrarian reform and rural development in Asia and the Far East' paper given at World Conference on Agrarian Reform and Rural Development, Rome: FAO.
Whyte, R. O. and Whyte, P. (1982) *The Women of Rural Asia*, Boulder, Colorado: Westview Press.
Wickramasekera, P. (1977) 'Aspects of the hired labour situation in rural Sri Lanka: some preliminary findings', in S. Hirashima (ed.) *Hired Labour in Rural Asia*, Tokyo: Institute of Development Economics.
Wickramasinghe, A. (1987a) *The Socio-economic Survey of Rural Women: the Innovation for Income Generating Activities for the Integrated Rural Development Project: Kegalle District*, Colombo: Ministry of Plan Implementation.
—— (1987b) 'Income generation for rural women in the Kegalle district under the integrated rural development project', paper given at the Department of Geography Staff Seminar Series, University of Peradeniya, Sri Lanka, June.
—— (1989a) *Status of Women in Rural Sri Lanka*, Colombo: Canadian International Development Agency.
—— (1989b) 'The participation of women in the labour force in crop production in the dry zone of Sri Lanka', paper given at the First National Convention on Women's Studies, Centre for Women's Research, March.

PART IV SOUTH-EAST ASIA AND OCEANIA

Introduction

Bryant, Jenny (1991) 'Housing and poverty in Fiji: who are the urban poor?', paper presented at the Pacific Science Congress, Waikiki, Honolulu, Hawaii, 27 May–2 June.
Chiang, Nora Lan-hung (1983) 'Female migration to Taipei: process and adaptation', *Population Geography*, 5 (1 and 2): 12–33.
—— (1984) 'The migration of rural women to Taipei', in J. T. Fawcett, *et al.* (eds) *Women in the Cities of Asia: Female Migration and Urban Adaptation*, Boulder, Colorado: Westview Press.
—— (1988) 'Changing role of women in Taiwan and its implications for development', Proceedings of the Second Asian-Pacific Cultural Symposium, Cultural and Social Centre for the Asian and Pacific Region, Seoul, 23–26 May.
Connell, John (1984) 'Status or subjugation? Women, migration and development in the South Pacific', *International Migration Review*, 18, 4, 964–83.
DesRochers, K. (1990) 'Women's fishing on Kosrae Island, Federated States of Micronesia: the effects of cultural, social and technological change on women's

use of the nearshore zone', unpublished master's thesis, Department of Geography, University of Hawaii, USA.
Elmhirst, Rebecca (1989) 'Women and work in Indonesia from 1970–1985: capitalism and gender inequality in perspective', unpublished master's thesis, Department of Geography, University of British Columbia, Canada.
Fahey, Stephanie (1979) 'Urban commitment: a case study from Bulolo', in R. Jackson, P. Batho and J. Ondago (eds) *Urbanisation and its Problems in Papua New Guinea*, Port Moresby: University of Papua New Guinea.
—— (1986) 'Development, labour relations and gender in Papua New Guinea', *Mankind*, 16, 2: 118–31.
—— (1988) 'Class, capital and spatial differentiation in Papua New Guinea', unpublished doctoral dissertation, Australian National University, Canberra.
Fairbairn-Dunlop, Peggy (1991) 'E au le inailau a tamaital': women, education, and development in Western Samoa, unpublished doctoral dissertation, Macquarie University, Sydney, Australia.
Larner, Wendy (1990) 'Samoan women's migration and work in New Zealand', unpublished Master's thesis, University of Waikato, New Zealand.
McGee, T. (1987) 'Mass markets–little markets: a call for research on the proletariarisation process, women workers and the creation of demand', in J. Momsen and J. Townsend (eds) *Geography and Gender in the Third World*, London: Hutchinson.
Preston, David (1986) 'Households and their livelihood strategies', Department of Human Geography, Research School of Pacific Studies, Australian National University, Canberra, mimeo.
—— (1989) Personal communication.
Sando, Ruth Ann (1981) 'Doing the work of two generations: the impact of depopulation on rural women in contemporary Taiwan', paper presented at the IGU Symposium on Rural Development, Fresno, April.
Singhanetra-Renard, Anchallee. (1984) 'Effect of female labour force participation on fertility: the case of construction workers in Chaing Mai City', in G. Jones (ed.) *Women in the Urban and Industrial Labour Force*, monograph 33, Canberra: Australian National University, Canberra.
—— (1987) 'Non-farm employment and female labour mobility in northern Thailand', in J. Momsen and J. Townsend (eds) *Geography of Gender in the Third World*, London: Hutchinson.
Thomas, Pam (1986) 'Women and development: a two-edged sword', in ACOA (eds) *Development in the Pacific: What Women Say*, ACOA Development Dossier 18, Canberra.
Thomas, Pam and Noumea, Simi (1983) 'The new Samoan businesswoman', *Pacific Perspective*, 11, 2: 5–12.
Young, Mei Ling and Kamal, Salih (1987) 'The Malay family: structural change and transformation – a research proposal', in Momsen and Townsend, *op. cit.*

14 Gender and the quality of life of households in raft-houses, Temerloh, Pahang, peninsular Malaysia

Azizah, Kassim, *et al.* (1986) 'Social amenities and the quality of life in squatter areas of Kuala Lumpur: some preliminary findings', paper presented at the seminar on Population and the Quality of Life, held in Malacca, 14–17 April.
Bubolz, M. M., Eicher, J. B. and Sontag, M. S. (1979) 'The human ecosystem: a model', *Journal of Home Economics*, 10, 28–31.
Bubolz, M. M., Eicher, J. B., Evers, S. J. and Sontag, M. S. (1980) 'A human

ecological approach to quality of life: conceptual framework and results of a preliminary study', *Social Indicators Research*, 7: 103–36.
Moller, V. and Schlemmer, L. (1983) 'Quality of life in South Africa: towards an instrument for the assessment of quality of life and basic needs', *Social Indicators Research*, 12: 225–79.
Sulaiman, Husna and Yahya, Nurizan (1986) 'Provision of housing and the quality of life of the urban poor', paper presented at the seminar on Population and the Quality of Life, held in Malacca, 14–17 April.
Whitehead, Ann (1985) 'Effects of technological change on rural women: a review of analysis', in Ahmed, Iftikhar (ed.) *Technology and Rural Women: Conceptual and Empirical Issues*, George Allen & Unwin.

15 Development and factory women

Ackerman, S. (1980) 'Cultural processes in Malaysian industrialization: a case study of Malay factory workers', Ph.D. thesis, San Diego: University of California.
Elson, D. and Pearson, R. (1981) 'The subordination of women and the internationalisation of factory production', in Young, K., Wolkowitz, C. and McCullagh, R. (eds) *Of Marriage and the Market*, London: CSE Books.
Fatimah, Daud (1985) *Minah Karan: the Truth about Malaysian Factory Girls*, Kuala Lumpur: Berita Publishing.
Government of Malaysia (1979) *Quarterly Survey of Employment in Manufacturing Industries, Peninsular Malaysia 1979*, Kuala Lumpur: Department of Statistics.
Grossman, R. (1979) 'Women's place in the integrated circuit', *Southwest Asia Chronicle*, 66.
HAWA (1981) *HAWA Draft Report*, University of Malaya, mimeo.
Jamilah, Ariffin (1980a) 'Industrial development in Peninsular Malaysia and rural urban migration of women workers: impact and implications', *Journal Ekonomi Malaysia*, 1: 1.
—— (1980b) 'The position of women workers in the manufacturing industries in Malaysia', in Consumer Association of Penang (eds) *Malaysian Women: Problems and Issues*, Penang: CAP Publishing.
—— (1982) 'Industrialization, female labour migration and the changing pattern of Malay women's labour force participation', *Journal of Southeast Asian Studies*, 10: 23–34.
—— (1984) 'Industrial development and rural–urban migration of Malay women workers in Peninsular Malaysia', Ph.D. thesis, Australia: University of Queensland.
Khoo, J. C. and Khoo, K. J. (1978) 'Work and consciousness: the case of electronics "runaways" in Malaysia', paper presented to the Conference on Continuing Subordination of Women in the Process of Development, Institute of Development Studies, Brighton.
Kraal, C. (1979) 'Small pay, crammed quarters', *New Straits Times*, 27 February, Kuala Lumpur.
Lim, L. Y. C. (1978) *Women in Multinational Corporations in Developing Countries: the Case of the Electronics Industry in Malaysia and Singapore*, occasional paper 9, Women's Studies Program, University of Michigan, Ann Arbor.
Yusuf, A. A. (1983) *The Holy Quran: Texts, Translations and Commentary*, Washington: Amana Corp.
Zainah, Anwar (1978) 'Roughing it out . . .', *New Sunday Times*, 12 March, Kuala Lumpur.

16 Women and agriculture in western Samoa

Aiono, Fanaafi (1986) 'Western Samoa: the sacred covenant' in Institute of Pacific Studies (ed.) *Land Rights of Pacific Women*, Suva: University of the South Pacific.

Brookfield, M and Ward, R.G. (eds) (1988) *New Directions in the South Pacific*, Academy of the Social Sciences in Australia, Research School of Pacific Studies, Canberra: Australia National University.

Fairbairn, Teo (1985) *Island Economies*, Suva: University of the South Pacific.

Holmes, J. (1974) *Samoan Village*, New York: Holt, Reinhart & Winston.

Salale, Salale (1976) *Economic Activity in Census Populations*, Analytical Report 2, Government of Western Samoa.

Schoeffel, P. (1980) 'O Tamaitai Samoa', Centre for Applied Studies in Development, Suva: University of the South Pacific.

—— (1984) 'Dilemmas of modernisation in primary health care in western Samoa', *Social Sc. Med.*, 19, 23: 209–16.

—— (1986) 'The rice pudding syndrome' in Development Dossier 18, *Development in the Pacific: What Women Say*, Suva: University of the South Pacific.

Sio, B. (1987) 'Nutrition and health', paper given at CTA/IRETA Conference on Appropriate Food Production Systems Around the Rural Homestead, Apia: University of the South Pacific.

Thomas, P. (1981) *The Missing Link*, Suva: University of the South Pacific.

—— (1986) 'Women and development: a two-edged sword' in Development Dossier 18, *Development in the Pacific: What Women Say*, Suva: University of the South Pacific.

—— (1987) 'Western Samoa: a population profile for the 80s', Canberra: Australia National University.

Thomas, P. and Hill, H. (1986) 'Major issues in nonformal education in the South Pacific', in D.C. Thorsby (ed.) *Human Resources Development in the Pacific*, Canberra: Australia National University, Pacific Policy Papers 3.

PART V LATIN AMERICA

Introduction

Calio, Sonia A. (1991) *Un Plan Directeur du Point de Vue des Femmes*, IGU Study Group On Gender and Geography, working paper 14, Department of Geography, University of Newcastle upon Tyne.

Franco, Maria Christina (1991) *Working Women and Urban space: Experiences in the South East of Bogota, Colombia*, IGU Study on Gender and Geography, working paper 22. Department of Geography, University of Newcastle upon Tyne.

Harry, Indra S. (1992) 'Women in agriculture in Trinidad: an overview', in Janet Momsen (ed.) *Women and Change: a Pan-Caribbean Perspective*, London: James Currey; Jamaica: Heinemann (Caribbean); and Bloomington: Indiana University Press.

Momsen, Janet H. (1986): Estudios de la mujer', in Colin Clarke, Silvana Levy *et al.* (eds) *Cambio social y economico en Latinamerica Perspectivas geograficas*, National University of Mexico.

Rossini, Rosa Ester (1989) 'Geografia e genero: a mulher na lavoura canaviera Paulista', doctoral dissertation, University of Sao Paulo, Brazil.

Sternbach, Nancy Saporta, Marysa Navarro-Aranguren, Patrica Chuchryk and

Sonia E. Alvarez (1992) 'Feminisms in Latin America: from Bogota to San Berbardo', *Signs* 17, 2: 393–434.
Tadeo, Nidia (1991) *Women's Work in Family-operated Commercial Flower Growing Units, Buenos Aires Metropolitan Area, Argentine*, IGU Study Group on Gender and Geography, working paper 18. Department of Geography, University of Newcastle upon Tyne.

17 The role of women's organisations and groups in community development

Agramont Virreira, M. *et al.* (1986) *Bibliografia de la mujer boliviana (1920–1985)*, La Paz: CIDEM.
Alzerreca, E. and Ruiz, C. B. (1987) *Palabra de Mujer: Dos Experiencias de Comunicacion*, La Paz: CIDEM.
Barrios de Chungara, D. with Viezzer, M. (1978) *Let Me Speak! Testimony of Domitila, a Woman of the Bolivian mines*, New York: Monthly Review Press.
Bronstein, A. (1982) *The Triple Struggle: Latin American Peasant Women*, London: WOW Campaigns Ltd.
CEDION (1989) 'A new democracy? The general elections 1989', *Bolivia Bulletin*, April, La Paz: CEDOIN
CIDEM, (1986) *Directorio de Instituciones Femininas*, La Paz: CIDEM.
Dunkerly, J. (1985) 'Back to the bad old ways', *The Guardian*, 14 December.
Figueroa, B. and Anderson, J. (1981) *Women in Peru*, London: Change.
Foronda, E. (1989) 'Number of street kids grow as Bolivian economy worsens', *Latinamerica Press*, Lima: 2 February.
Gott, R. (1987) 'The bitter harvest', *The Guardian*, 9 October.
Harris, O. (ed.) (1983) *Latin American Women*, London: Minority Rights Group.
Latin American and Caribbean Women's Collective (1980) *Slaves of Slaves: the Challenge of Latin American Women*, London: Zed Press.
Midgley, J., Hall, A., Hardiman, M. and Narine, D. (1986) *Community Participation, Social Development and the State*, London: Methuen.
O'Kelly, E. (1978) *Rural Women: Their Integration in Development Programmes and How Simple Intermediate Technologies Can Help Them*, London: O'Kelly.
Organisation for Economic Co-operation and Development (1988) *Voluntary Aid for Development: the Role of Non-Governmental Organisations*, Paris: OECD.
Salmen, L. F. (1988) 'Determining success in development', *Grassroots Development*, Inter-American Foundation, 12: 3.
Sandoval, Z. G. (1987) *Organizaciones No Gubernamantales de Desarrolla (Planificacion y Evaluacion)*, La Paz: SENPAS and UNITAS.
Seager, J. and Olson, A. (1986) *Women in the World: an International Atlas*, London: Pan Books.
Swift, R. (1988) 'From the grassroots: solving the development puzzle', *The New Internationalist*, May.
Vacaflor, H. (1988) 'Pricing inflation out of the market', *South*, August, 28.
Van Wijk-Sijbesma, C. (1986) 'Helping women to help themselves', *Waterlines*, April.
World University Service (undated) *WUS Women, Education and Development Campaign*, London: WUS.

18 Deconstructing the household

Bernstein, H. (1979) 'African peasantries: a theoretical framework', *Journal of Peasant Studies*, 6, 4: 421–43.
Bourque, S. and Warren, K. (1981) *Women of the Andes: Patriarchy and Social Change in Two Peruvian Towns*, Ann Arbor: University of Michigan Press.
Collins, J. (1983) 'Fertility determinants in a high Andes community', *Population and Development Review*, 19, 1: 61–75.
Dandler, J. (1971) 'Politics of brokerage, leadership and patronage in the *campesino* movement of Cochabamba, Bolivia (1932–1952)', unpublished Ph.D. thesis, Madison: University of Wisconsin.
Deere, C.D. (1978) 'The differentiation of the peasantry and family structure: a Peruvian case study', *Journal of Family History*, 3, 4: 422–38.
Ellis, F. (1988) *Peasant Economics: Farm Household and Agrarian Development*, Cambridge: Cambridge University Press.
Friedmann, H. (1980) 'Household production and the national economy: concepts for the analysis of agrarian formations', *Journal of Peasant Studies*, 7, 2: 158–84.
—— (1986) 'Patriarchal commodity production', in A. MacEwan Scott (ed.) *Rethinking Petty Commodity Production*, Social Analysis Special Issue 20: 47–55.
Harris, O. (1981) 'Households as natural units', in K. Young, C. Wolkowitz and R. McCullagh (eds) *Of Marriage and the Market*, London: CSE Books.
Long, N. (1977) 'Commerce and kinship in the Peruvian highlands', in R. Bolton and E. Meyer (eds) *Andean Kinship and Marriage*, Washington: American Anthropological Special Publication 7.
—— (1986) 'Commoditisation: thesis and antithesis', in N. Long *et al.* (eds) *The Commoditisation Debate: Labour Process, Strategy and Social Network*, Wageningen: Agricultural University Department of Sociology Papers 17.
Pearson, M. (1987) 'Old wives or young midwives? Women as caretakers of health: the case of Nepal', in J. H. Momsen and J. Townsend (eds) *Geography of Gender in the Third World*, London: Hutchinson.
Radcliffe, S. (1986) 'Gender relations, peasant livelihood strategies and migration: a case study from Cuzco, Peru', *Bulletin of Latin American Research*, 5, 2: 29–47.
Redclift, N. (1985) 'The contested domain: gender, accumulation and the labour process', in N. Redclift and E. Mingione (eds) *Beyond Employment: Household, Gender and Subsistence*, Oxford: Basil Blackwell.
Smith, C. (1986) 'Reconstructing the elements of petty commodity production', in A. MacEwan Scott (ed.) *Rethinking Petty Commodity Production*, Social Analysis Special Issue Series 20: 29–46.
Whitehead, A. (1981) '"I'm hungry mum': the politics of domestic budgeting', in K. Young, C. Wolkowitz and R. McCullagh (eds) *Of Marriage and the Market*, London: CSE Books.
—— (1984) 'Women's solidarity and divisions among women', *Bulletin of the Institute of Development Studies*, 15, 1: 6–11.

19 Women's roles in colonisation

Caceres, Ingrid (1980) 'La division de trabajo por sexo en la unidad campesina minifundista', in M. Leon, *Mujer y capitalismo agrario*, Bogota: ACEP.
Castro Caycedo, G. (1976) *Colombia Amarga*, Bogota: Carlos Valencia.
CIPAF (1985) *Encuesta nacional de la mujer rural*, Santa Domingo: CIPAF.

Deere, Carmen Diana and Leon, Magdalena (1982) 'Produccion campesina, proletarization y la division sexual del trabajo en la Zona Andina', in M. Leon, *Las trabajadoras del agro: debate sobre la mujer en America Latina y el Caribe*, Bogota: ACEP.
—— (1983) 'Women in Andean agriculture' in L. Beneria (ed.) *Women and Development*, New York: Praeger.
Gonzalez, C. (1980) 'Formacion y cambios del latifundio ganadero y sus efectos en la organizacion de la familia campesina', in M. Leon, *Muyer y capitalismo agrario*, Bogota: ACEP.
Hamilton, S. (1986) *An Unsettling Experience: Women's Migration to the San Julian Settlement Project*, New York: Institute for development anthropology, Co-operative Agreement for Human Settlement and Natural Resource Systems Analysis.
Hecht, Susan (1986) *Informe de consultoria*, Bogota: Corporacio de Araracuara.
—— (undated) 'The Latin American livestock sector and its potential impacts on women', mimeo.
Lisansky, J. (1979) 'Women in the Brazilian frontier', *Latinamericanist*, 15: 1.
Molano, A. (1987) *Selva adentro: Una historia oral de la colonizacion del Guaviare*, Bogota: El Ancora.
Santamaria, G. (1978) 'La nina de la selva', *El Tiempo*, 30–1 July.
Townsend, J. and Wilson de Acosta, S. (1987) 'Gender roles in colonisation of rainforest', in J. Momsen and J. Townsend (eds) *Geography of Gender in the Third World*, London: Hutchinson.

20 Housewifisation and colonisation in the Colombian Rainforest

Burfisher, M. and Horenstein, N. (1985) *Sex Roles in the Nigerian Tiv Household*, West Hartford: Kumarian Press.
Deere, C. D. (1986) 'La mujer rural y la politica estatal: La experiencia Latinoamericana y Caribena de reforma agraria', in M. Leon and C. D. Deere (eds) *La mujer y la politica agraria en America Latina*, Bogota: Siglo Veintiuno.
Deere. C. D. and Leon de Leal, M. (1983) *Women in Andean Agriculture*, Geneva: ILO.
Dixon-Mueller, R. (1985) *Women's Work in Third World Agriculture*, Geneva: ILO.
Gill, L. (1987) *Peasants, Entrepreneurs and Social Change: Frontier Development in Lowland Bolivia*, Boulder: Westview.
Gonzalez, C. (1980) 'Formacion y cambios del latifundio ganadero y sus efectos en la organizacion de la familia campesina' in M. Leon de Leal (ed.) *Mujer y capitalismo agrario*, Bogota: ACEP.
Hamilton, S. (1986) *An Unsettling Experience: Women's Migration to the San Julian Colonisation Project*, Institute for Development Anthropology Inc., paper 26.
Hanger, J. and Moris, J. (1967) 'Women and the household economy', in R. Chambers and J. Moris (eds) *Mwea: an Irrigated Rice Settlement in Kenya*, Munchen: Weltforum Verlang, Afrika Studien, 38.
Hecht, S. B. (1985) 'The Latin American livestock sector and its potential impacts on women', in J. Monson and M. Kalb (eds) *Women as Food Producers in Developing Countries*, Los Angeles: University of California Press.
Jacobs, S. (1984) 'Women and land resettlement in Zimbabwe', *Review of African Political Economy*, 27–8: 33–50.
Leon, M. (1986) 'Politica agraria en Colombia y debate sobre politicas para la mujer rural', in M. Leon and C. D. Deere (eds) *La mujer y la politica agraria en America Latina*, Bogota: Siglo Veintiuno.

—— (1987) 'Politicas agrarias en Colombia y discusion sobre la politica para la mujer campesina', *Estudios Rurales Latinoamericanos*, 10,1: 71–93.
Leon, M. and Prieto, P. and Salazar, M. C. (1987) *Acceso de la mujer a la tierra en america latina y el caribe, panorama general y estudios de case de Honduras y Colombia*, Bogota: Informe presentado a FAO.
Lisansky, J. (1979) 'Women in the Brazilian frontier', *Latinamericanist*, 15,1: 1–3.
—— (1981) 'Santa Terezinha: Life in a Brazilian frontier town', Ph.D. thesis, University of Florida.
Massey, D. (1984) *Spatial Divisions of Labour: Social Structures and the Geography of Production*, London and Basingstoke: Macmillan.
Meertens, D. (1988) 'Mujer y colonizacion en el Guaviare (Colombia)', *Colombia Amazonica*, 3, 2: 21–59.
—— (1989) 'Women and colonisation in Latin America: reflections around a case study in Colombia', paper presented at the Commonwealth Geographical Bureau Workshop on Gender and Development, Newcastle upon Tyne.
Mies, M. (1986) *Patriarchy and Accumulation on a World Scale: Women in the International Division of Labour*, London: Zed.
Moser, C. O. N. (1987) 'Women, human settlements and housing: a conceptual framework for analysis and policy making', in C. Moser and L. Peake (eds) *Women, Human Settlements and Housing*, London: Tavistock.
Redclift, N. (1985) 'The contested domain: gender, accumulation and the labour process', in N. Redclift and M. Mingione (eds) *Beyond Employment: Household, Gender and Subsistence*, Oxford: Basil Blackwell.
Townsend, J. (1988) 'Co-operativa integral de la Comuna de Payoa', *Cuadernos de Agroindustria y Economia Rural*, 20: 51–70.
Townsend, J. and Wilson de Acosta, S. (1987) 'Gender roles in the colonisation of rainforest: a Colombian case study', in J. Momsen and J. Townsend (eds) *Geography of Gender in the Third World*, London: Hutchinson.
Ulluwishewa, R. (1989) 'Development planning and gender inequality: a case study in Mahaweli Development Project, Sri Lanka', paper presented at the Commonwealth Geographical Bureau Workshop on Gender and Development, Newcastle upon Tyne.
Valestrand, H. (1989) 'Making peasant women visible: a sufficient research strategy with a case study of Coto Sur, Costa Rica', paper given at the Commonwealth Geographical Bureau Workshop on Gender and Development, Newcastle upon Tyne.
Walby, S. (1986) *Patriarchy at Work*, Cambridge: Polity Press.
Weil, J. (1980) 'The organization of space in a Quechua pioneer settlement', Ph.D. thesis, Ann Arbor: University of Michigan.

21 The role of gender in peasant migration

Behrman, Jere and Wolfe, Barbara (1984) 'Micro-determinants of female migration in a developing country: labour market, demographic marriage market and economic marriage market incentives', *Research in Population Economics*, 5: 137–66.
Bernstein, H. (1982) 'Notes on capital and the peasantry', in Harriss, J. (ed.) *Rural Development: Theories of Peasant Economies and Agrarian Change*, London: Hutchinson.
Boserup, E. (1970) *Women's Role in Economic Development*, London: Allen & Unwin.

REFERENCES

Bourdieu, P. (1977) *Outline of a Theory of Practice*, Cambridge: Cambridge University Press.

Bourque, S. and Warren, K. (1981) *Women of the Andes: Patriarchy and Social Change in Two Peruvian Towns*, Ann Arbor: University of Michigan Press.

Collins, J. (1986) 'The household and relations of production in southern Peru', *Comparative Studies in Society and History*, 28, 4: 651–71.

—— (1988) *Unseasonal Migrations: the Effects of Labor Scarcity in Peru*, Princeton: Princeton University Press.

Deere, C.D. (1978) 'The development of capitalism in agriculture and the division of labour by sex: a study of the Northern Peruvian Sierra', Ph.D. thesis, University of California, Berkeley.

Deere, C.D. and Leon de Leal, M. (1982) *Women in Andean Agriculture*, Geneva: ILO.

Gonzales de Olarte, E. (1984) *Economia de la comunidad campesina*, Lima: Instituto de Estudios Peruanos.

Goodman, D. and Redclift, M. (eds) (1981) *From Peasant to Proletarian: Capitalist Development and Agrarian Transitions*, Oxford: Basil Blackwell.

International Migration Review (1984) Special issue on women and migration, 18 (winter).

ISIS (1980) Issue on women and migration in Latin America, 14.

Khoo, S., Smith, P. and Fawcett, J. (1984) 'Migration of women to cities: the Asian situation in comparative perspective', *International Migration Review*, 18: 1247–63.

Laite, J. (1985) 'Circulatory migration and social differentiation in the Andes' in Standing, G. (ed.) *Labour Circulation and the Labour Process*, London: Croom Helm.

Lehmann, D. (ed.) (1982) *Ecology and Change in the Andes*, Cambridge: Cambridge Univesity Press.

Migration Today (1982) Special issue on women and migration, 10, 3–4.

Orlansky, D. and Dubrovsky, S. (1978) *The Effects of Rural-Urban Migration on Women's Role and Status in Latin America*, Paris: UNESCO.

Pittin, R. (1984) 'Migration of women in Nigeria: the Hausa case', *International Migration Review*, 18, 4: 1293–1315.

Quijandria, B. and Espinoza, C. (1988) 'Sistemas de produccion y economia campesina: caracterizacion y estrategias productivas como base de politicas agrarias', in Eguren, F., Hopkins, R., Kervyn, B., and Montoya, R. (eds) *Peru: el problema agrario en debate*, Lima: SEPIA II.

Radcliffe, S. (1986a) 'Women's lives and peasant livelihood strategies. A study of migration in the Peruvian Andes', Ph.D. thesis, University of Liverpool, Liverpool.

—— (1986b) 'Gender relations, peasant livelihood strategies and mobility: a case study from Cuzco, Peru', *Bulletin of Latin American Research*, 5, 2: 29–47.

Radcliffe, S. (1987) 'Chacrawarmi or Ilactawarmi: migrant women and the circulation of goods in the Peruvian Andes', I Simposio de America Latina, Warsaw, 1987.

Schmink, M. (1984) 'Household economic strategies: review and research agenda', *Latin American Research Review*, 19, 3: 87–102.

Thadani, V. and Todaro, M. (1979) *Female Migration in Developing Countries: a Framework for Analysis*, Centre for Policy Studies, New York: Population Council.

Todaro, M. (1985) *Economic Development in the Third World*, 3rd edn, Longman: New York and London.

Trager, L. (1984) 'Family strategy and the migration of women: migration to Dagupan City, Philippines', *International Migration Review*, 18, 4: 1264–78.
Wilkinson, R.C. (1983) 'Migration in Lesotho: some comparative aspects with particular reference to the role of women', *Geography*, 68: 3.
Young, K. (1979) 'The continuing subordination of women in the development process', *IDS Bulletin*, 10, 3: 14–37.
Young, K. (1980) 'The creation of a relative surplus population: a case study from southern Mexico', *SLAS Bulletin*, 32.
Youssef, N. (ed.) (1979) *Women in Migration: a Third World Focus*, Washington, DC: USAID.

INDEX